Statistics and Computing

Statistics and Computing (SC) includes monographs and advanced texts on statistical computing and statistical packages.

Antonio Zinilli

Elements of Network Science

Theory, Methods and Applications in Stata, R and Python

Springer

Antonio Zinilli
Research Institute on Sustainable Economic Growth
National Research Council
Rome, Italy

ISSN 1431-8784 ISSN 2197-1706 (electronic)
Statistics and Computing
ISBN 978-3-031-84711-0 ISBN 978-3-031-84712-7 (eBook)
https://doi.org/10.1007/978-3-031-84712-7

This Springer imprint is published by the registered company Springer Nature Switzerland AG
The registered company address is: Gewerbestrasse 11, 6330 Cham, Switzerland

For Claudia, Valentina, and my family—you are my life, my love, my light.

Preface

Network Science is the study of complex systems composed of basic components known as nodes and their interactions or relationships, referred to as edges (Newman, 2003). By studying an interacting system as a whole through this mathematical abstraction, we can uncover important insights that would otherwise be overlooked if we considered the system merely as a collection of isolated units. Network Science provides a unified formalism that can be applied to describe systems from different scientific fields while still being represented as a network of interconnected elements. Indeed, using this all-inclusive approach, a variety of complex systems emerging from various disciplines have been studied.

The book provides readers with both theoretical and applied tools for a correct implementation of Network Science (NS) techniques. Network Science offers a means to understand and analyze complex systems that involve various types of relationships. These systems can include transportation networks, computer networks, and social and economic networks. For instance, the structure and dynamics of relationships within social and economic contexts play a crucial role in shaping human behavior, decision-making, and influencing the outcomes of related processes. In the social sciences, Network Science can be employed to examine the dissemination of information, ideas, and influence. It also helps in understanding how social networks impact health behaviors, political opinions, and cultural norms. In economics, Network Science can be used to investigate the structure and evolution of markets, innovation processes, and financial systems. It helps to understand how network structure influences firm behavior, market efficiency, and the distribution of wealth and power. Additionally, it helps to understand how network structure influences firm behavior, market efficiency, and the distribution of wealth and power.

As such, the book provides readers with a comprehensive toolbox for designing rigorous and effective Network Science using three popular statistical software platforms: Stata, R, and Python. Theoretical statistical models pertaining to each individual network technique are discussed, followed by empirical applications that use both built-in and user-written packages. The objective is to present the mathematical and statistical foundations of each model, along with demonstrations involving hand calculations (when applicable) and step-by-step code implementation.

Additionally, each chapter has at least one case study that illustrates one of the several techniques discussed. In essence, this book bridges the gap between theoretical understanding and practical application, making Network Science more accessible to a wide range of users. The intended audience for this book includes beginning and advanced postgraduate students, researchers, and data scientists who are interested in using Network Science techniques, regardless of the software platform they use.

The inclusion of applications in Stata, R, and Python aims to provide readers with a more comprehensive understanding of the availability of Network Science (NS) methods across different software platforms. Since these three software platforms use distinctive application processes, this book will guide the reader in navigating the different coding techniques adopted in each software, supplemented by specific case studies.

The first chapter serves as a road map for readers, providing an initial understanding of the contexts in which networks arise, the scientific questions being addressed, and the measurements being used. Additionally, this section offers a quick overview of how to install all the required packages (in Stata, R, and Python) that will be used throughout the book.

The second chapter focuses on the main concepts related to network properties. Practical cases using raw and unformatted data are illustrated, aiming to present powerful code snippets with step-by-step explanations in Stata, R, and Python. These examples cover the management of edge lists and adjacency matrices, as well as specific cases of one- and two-mode networks. The chapter concludes with the network visualization and the implementation of textual networks using associated scripts in the three software platforms. A case study in this chapter demonstrates how to build and analyze a textual network using the RStudio environment.

The aim of the third chapter is to introduce the primary definitions and concepts in Network Science. These are essential for understanding the fundamentals of graph theory and the subsequent micro, meso, and macro measurements. For each measure in this chapter, statistical fundamentals, manual calculations, step-by-step code implementation in the three software platforms, and real data examples are provided.

In the fourth chapter, various classes of graphs observed in real contexts are presented. The power law distribution and other typical statistical distributions are discussed initially. Following that, an overview of network properties is provided, along with discussions on concepts such as small-world properties and clustering procedures.

The fifth chapter explores key models in social network analysis, including the QAP model and logistic regression. Furthermore, Stochastic Actor Oriented Models (SAOM) and Exponential-Family Models of Random Graphs (ERGM, TERGM, and GERGM) are discussed for modeling both static and dynamic random graphs with endogenous and exogenous attributes. This involves considering social relationships as the minimum unit of observation and studying endogenous characteristics and individual or exogenous attributes (e.g., gender, age, socio-economic status) to explain the relational configurations that constitute the overall structure.

Each chapter includes real-world data applications, encompassing text networks, knowledge networks (co-publication and citation), innovation networks (co-patenting), and friendship networks. All the code and datasets used in the case studies presented in this book, located at the end of each chapter, are available for download through GitHub at the following path: https://github.com/AntZin84/NetworkBook.git.

Rome, Italy Antonio Zinilli

Acknowledgments

First and foremost, I extend my profound gratitude to my colleagues at CNR-IRCrES. Their wisdom, insightful guidance, and unwavering support over the years have been invaluable to my journey. I am particularly indebted to Emanuela Reale and Giovanni Cerulli, whose incisive insights and relentless efforts have significantly enriched this work. Special thanks go to Martina Neuländtner, Arash Hajikhani, Luca Andriani, and Emanuela Varinetti for reading specific chapters of this book. All errors remain mine.

Throughout this challenging academic journey, my wife Valentina's steadfast support, constant encouragement, and remarkable patience have been my guiding light. She has walked with me every step of the way, playing an essential role in helping me balance work and family life.

In conclusion, I wish to express my deepest gratitude to my parents, Giuseppe and Silvana, whose unconditional love and invaluable guidance have been the foundation of my personal and professional growth. I am also profoundly thankful to my brother, Pierfrancesco, for his constant support. Their enduring influence has been instrumental in shaping me into the person I am today.

To all who have contributed to its fruition, my heartfelt thanks.

Acknowledgements

Contents

Chapter 1
Introduction

1.1 Intent of This Book

Network Science is a relatively new branch of complexity science that brings together diverse disciplines such as physics, sociology, biology, and computer science. The expansion of Network Science in various disciplinary areas, which have recognized its significance and embraced its mission, can be attributed to the integration of computer science, sociological theory, and mathematics. The widespread adoption of Network Science techniques has validated their applicability in studying a broad range of socio-economic phenomena and processes. This is facilitated by the increasing availability of tools for automated processing of network data and the richness of large datasets. In recent years, the diffusion of techniques and procedures developed in the field of Network Science has reached a stage of gradual consolidation. Most real-world socio-economic systems can be effectively represented using network theory, where nodes represent entities and edges represent their connections.

Within the pages of this book, our aim is to construct a bridge between the long-standing tradition of Social Network Analysis (SNA) in the mathematical social sciences and the field of Network Science (NS) from statistical mechanics (also called statistical physics). It is crucial to clarify why the choice was made to refer to it as Network Science (NS) rather than Social Network Analysis (SNA) in the title from the beginning. In the following explanation, we will provide the reasons behind this decision.

From Moreno's sociograms (1934, 1947) to White's kinship models (1963), Milgram's "Six Degrees of Separation" (1967a), Lévy-Strauss' structural anthropology (1974), and Granovetter's "Strength of Weak Ties" (1973), the concept of networks has played a prominent role in various schools of thought within the social sciences. The field of Social Network Analysis (SNA) emerged from these extensive explorations of the relational nature of social phenomena. SNA examines the

A. Zinilli, *Elements of Network Science*, Statistics and Computing,
https://doi.org/10.1007/978-3-031-84712-7_1

structure of a single network in relation to a specific social science theory. For example, it considers debates on whether a triad is a significant social unit and theories on the overall structure of groups, organizations, or societies in sociology (Freeman, 2004; Freeman et al., 1989; Carrington et al., 2005). Consequently, SNA techniques and procedures are strongly influenced by the social theory under investigation. SNA investigates the structure of interactions among multiple entities, such as individuals, enterprises, and organizations, as well as the interdependencies in behavior or attitudes related to social relationship configurations (Zinilli, 2016). Building on this rich history, the discipline continues to debate the empirical nature of the networks it studies and faces methodological challenges associated with its tools. For instance, Exponential Random Graph Models (ERGMs) discussed in Chap. 5 have emerged as valuable tools within Social Network Analysis (SNA). These models allow researchers to test hypotheses about the formation and structure of social networks, providing insights into the underlying mechanisms driving network formation (Filippetti & Zinilli, 2023).

Network Science emerged as a highly interdisciplinary research area focused on developing theoretical and practical approaches and techniques (Börner et al., 2007). The field often traces its origins to graph theory, particularly the random graph model proposed by Erdős and Rényi (1960). Network Science, characterized by its emphasis on complex networks, encompasses a broader and more structured scope than SNA. It aims to identify universal structural features in diverse complex networks, understand the mechanisms underlying their formation, and explore the relationship between features and network processes. Analogies from statistical physics are frequently employed to address challenges and problems in Network Science, as they offer phenomenon-oriented perspectives. The field seeks models that are theoretically grounded and capable of replicating the network's global or systemic characteristics by revealing insights into the data generation process. The models delineated in Network Science exhibit broad scalability compared to those in Social Network Analysis. Scalability, in terms of scale invariance, refers to the property of a model to maintain consistent performance regardless of changes in the size or complexity of the data. For instance, studies on scale-free networks or preferential attachment mechanisms, which can explain the power-law distribution, have received considerable attention in Network Science (Newman, 2003).

In summary, Social Network Analysis (SNA) is a subfield of sociology that focuses on studying social networks and the social structures that emerge from them. While SNA incorporates methods from mathematics and other disciplines, its main emphasis is on understanding social phenomena, such as social influence, information diffusion, and the formation and evolution of social interactions. On the other hand, Network Science is a broader and more interdisciplinary field that investigates networks across various domains beyond just social networks. The book will use the term Network Science rather than Social Network Analysis because it aims to provide a broader and more inclusive perspective on network research, especially when it comes to considerations on a larger scale and interdisciplinarity. It will delve into models influenced not only by social theory (e.g., homophily) but also by social network models (e.g., exponential random graph models), network formation methods primarily derived from works in statistical mechanics (e.g., scale-free networks), and computational techniques from computer science.

To analyze the social world through its network structure, Network Science is approached from a theoretical-methodological-computational standpoint. This integration from diverse disciplines allows us to begin analyzing the complex structure of society, which is composed of multiple interdependencies and interconnectedness.

1.2 The Seven Bridges of Königsberg

Very few scientific disciplines can be traced back to a specific point in history. However, Network Science, particularly its mathematical foundation known as graph theory, can make such a claim. Its origins can be traced back to 1735 in Königsberg (now Kaliningrad), located in East Prussia, which is presently a Russian enclave between Poland and Lithuania. Königsberg is situated at the estuary of the Pregel River, where two islands are formed.

During the eighteenth century, the city's various sections were interconnected by a system of seven bridges (see Fig. 1.1). It was customary for aristocratic families of that era to take walks along the riverbanks and the islands. This unique layout presented a modern enigma: is it possible to traverse all seven bridges without crossing the same one twice? Despite numerous attempts, no one succeeded in finding such a path. The mystery remained unsolved until 1735 when Leonard Euler, a Swiss-born mathematician, provided a robust mathematical proof demonstrating the impossibility of such a path (Euler, 1741).

The specific locations of the four regions *A*, *B*, *C*, and *D* are irrelevant to this problem. The only crucial aspect is the connections between these regions, which are denoted by the seven points *a*, *b*, *c*, *d*, *e*, *f*, and *g*. By removing unnecessary details, Euler devised a symbolic illustration. In this representation, the vertices

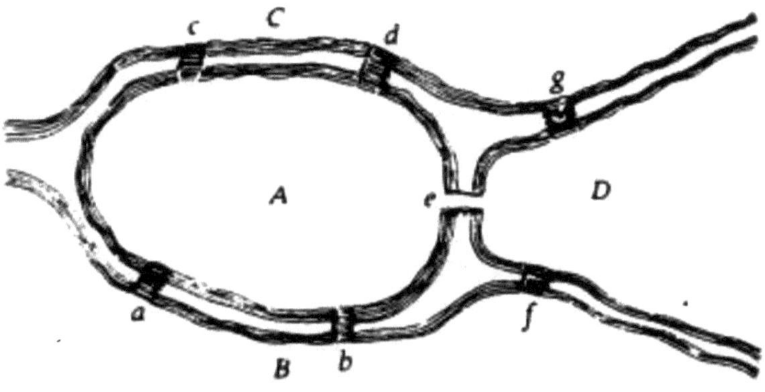

Fig. 1.1 The Bridges of Königsberg

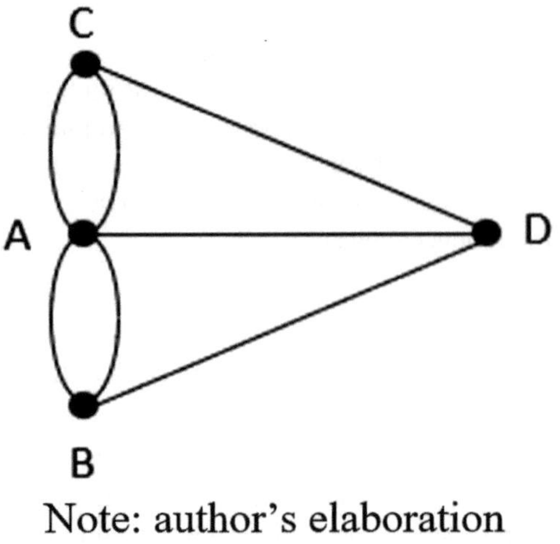

Fig. 1.2 Configuration of the Euler graph. Note: author's elaboration

(nodes) correspond to the four areas, while the arcs (edges, links, or ties) represent the connections formed by the seven bridges between the different areas. As a result, Euler constructed a graph, where the nodes represent sections of land and the edges represent the bridges (Fig. 1.2).

Euler made a simple observation: if a path traverses all bridges without repeating any bridge, nodes with an odd number of connections must serve as either the starting or ending point of the path. In fact, if you encounter a node with an odd number of connections, you will find that there is no remaining unused connection for you to continue. As a consequence, the number of edges incident to any vertex, except for the initial and final ones, must be an even number. As the Königsberg graph had four nodes with an odd number of connections (*A*, *B*, *C*, and *D*), no solution could be found for the problem. Euler's proof marked the first instance of resolving a mathematical problem using a graph.

The Bridges of Königsberg problem demonstrates that representing a problem as a graph simplifies it and makes it more manageable. Moreover, in many cases, the solution is derived from properties inherent to the graph itself. In other words, networks have inherent structural features that can either constrain or enable specific processes. In the following paragraph, we will provide a precise definition of the terms "graph" or "network" and delve deeper into some of their properties. Throughout this book, the terms "graph" and "network" are used interchangeably to convey the same type of information.

1.3 Fundamental Properties of Real Networks

In the centuries following Euler's exploration of the Königsberg bridges, graph theory gained popularity in mathematics. A significant milestone occurred in 1959 when Paul Erdős and Alfréd Rényi, two mathematicians and friends, proposed an elegant mathematical model that seemed capable of explaining complex phenomena. This marked the beginning of the study of networks, which are characterized by numerous interconnected components, and it gained unprecedented importance. Erdős and Rényi (1960) focused their research on the processes that generate networks. By using the concept of randomness to address questions about graph formation, their work pushed the study of networks to new heights.

In a random network, connections between nodes occur in a random order, implying that each node has an equal chance of acquiring an edge. Real networks, however, despite their complexity, exhibit internal order. Random networks exhibit characteristics of a Small World: highly connected networks where it is possible to reach any node from another in just a few steps. The 'Small World' concept originates from the idea that any two individuals can be connected through a short chain of intermediaries—typically no more than five. Initially defined in the social sector, the concept of "six degrees of separation" was introduced by Hungarian novelist Frigyes Karinthy (1929) and expanded upon by sociologist Stanley Milgram (1967a, 1967b). We now understand that this property is not exclusive to social networks but rather a common feature in most real networks within complex systems. This property endows networks with remarkable efficiency in terms of information transmission speed (Watts, 2003; Biagini et al., 2019).

Another characteristic observed in real networks is the tendency to cluster, with communities consisting of individuals who are closely related to one another. This property, known as transitivity in sociology, essentially measures the extent to which an individual's friends are also friends with each other (Wasserman & Faust, 1994). Consequently, clustering represents a key feature observed in various real networks.

The sociologist Mark Granovetter's influential work titled "The Strength of Weak Ties" (1973) played a crucial role in understanding how to combine the properties of small world and clustering. Granovetter highlighted the paradox that weak ties are the most crucial iin ensuring connectivity across the network. They serve as the essential links between closely related groups of people. This insight was later embraced by Duncan Watts and Steven Strogatz (1998), who built upon Granovetter's work and realized that the presence of these weak ties between separate yet internally cohesive communities guarantees the small-world effect, while "strong" ties ensure local aggregation.

Building on the model proposed by Watts and Strogatz (1998), studies of real networks have unveiled unexpected properties absent in previous network models. The distribution of node degrees in these networks follows a power-law instead of a Poissonian pattern. This results in the coexistence of numerous nodes with few connections and a small number of highly connected nodes, with the power law holding true regardless of the network's scale or the number of observed nodes. This discovery led to the development of scale-free network models, most notably the Barabási–Albert model proposed by Barabási and Albert (1999). This model explains network formation through successive node aggregation, assuming that new nodes are more likely to connect to already highly connected nodes ("preferential attachment"). Mathematically, this mechanism manifests in the power distribution of node degrees in networks. Scale-free networks exhibit a topology dominated by a small number of highly important and well-connected nodes known as "hubs" which connect the less well-connected nodes within the system. These properties have sparked interest in researching the topological elements of social and economic networks, particularly regarding more dynamic processes.

The field of Network Science has identified specific characteristics in real networks through hundreds of studies, unifying seemingly disparate systems, both natural and social. This unification allows the analysis of various phenomena using the same conceptual framework. Hidden similarities exist between seemingly unrelated systems such as bird flocks, the Internet, and metabolic networks, as they share the same type of interaction structure. The topology of the network strongly influences collective behavior. Therefore, Network Theory focuses on the study of the topological properties of these objects.

1.4 Several Approaches in Network Studies

The study of networks encompasses several disciplines, including sociology, economics, computer science, and physics, each bringing unique perspectives and methodologies to the field. Methodologically, network models are primarily explored by physicists and social scientists (such as statisticians and sociologists), although they approach the subject from distinct viewpoints and employ different methods.

Physicists investigate networks using tools from statistical mechanics and graph theory. Their focus often involves analyzing the specific properties of networks (Borgatti et al., 2009). For instance, they examine the degree distribution (the number of connections per node) and analyze the distances between nodes, focusing on the lengths of shortest paths between node pairs. Additionally, they investigate network dynamics, examining how networks evolve over time and how information spreads through them. Physicists aim to identify universal patterns and values governing network behavior across domains.

Social scientists, on the other hand, concentrate on generating statistical models to analyze social networks. They frequently employ probabilistic models to extract information from network data, such as identifying important nodes, detecting clusters or communities, and characterizing the relationship between network structure and other variables of interest. Social scientists are particularly interested in developing methods to analyze complex network data, including situations involving missing values or multiple types of links (Newman, 2003; Huisman, 2009).

Another distinction between physicists and social scientists in the study of networks lies in their emphasis on practical applications. Physicists tend to focus on theoretical network models and the underlying mechanisms governing their behavior, while statisticians prioritize creating practical applications for network analysis. Furthermore, they differ in the type of data they typically handle. Physicists sometimes work with highly idealized network models, whereas statisticians primarily deal with noisy or incomplete real-world data. This requires that statisticians develop methods to address data quality issues and draw conclusions based on incomplete or biased data.

Lastly, physicists and social scientists may approach the interpretation of their findings differently. Physicists often seek universal patterns and mechanisms that apply to a broad range of network types, while social scientists focus on drawing conclusions specific to individual networks. However, the approaches of social scientists often lack scalability, making them less suitable for the extensive big data networks commonly analyzed in Network Science. For example, ERGM models face computational challenges when dealing with large datasets. While physicists and social scientists approach network research from different perspectives, their work complements each other and contributes to a better understanding of the complex structures underlying various natural and social phenomena.

This book adopts an integrative approach, aiming to bridge the tradition of analyzing social networks in the mathematical social sciences (SNA) with Network Science in Physics (NS).

1.5 Advancements in Network Research

In recent years, Network Science has experienced considerable growth in its application to fields such as economics and social sciences. This development can be attributed largely to the increasing availability of data, particularly in the domain of big data. The advent of the big data era has developed network analysis in social and economic contexts. Over time, network datasets have undergone an exponential surge, transitioning from small networks with a few nodes to massive networks comprising millions, or even billions, of nodes. The digitization of data in the 1990s marked a key moment, enabling more efficient storage, manipulation, and analysis of extensive datasets by converting analog data into digital form. However, the true explosion of network data occurred with the rise of the Internet and the widespread proliferation of various large-scale networks. These networks spanned diverse domains such as physical infrastructures, the World Wide Web, and online social networks, leading to the exponential growth of available network data. The widespread accessibility of the Internet gave rise to numerous online platforms and social networks, significantly amplifying the scale and complexity of available data. This proliferation of large network datasets has posed both challenges and opportunities for social scientists.

These datasets encompass a wide range of networks, including knowledge and innovation networks where nodes represent researchers, universities, firms, or regions, and links represent collaborations or citations. Trade networks consist of nodes representing countries or regions, with links indicating import/export flows. Migration networks involve regions or countries as nodes and links representing connections between origin and destination regions or countries. Additionally, the network of businesses includes nodes representing firms and links representing trades between them. Analyzing these networks, even those with massive dimensions, allows researchers to determine the degree of network connectivity, the distance between nodes, and the presence of clusters or communities (as exemplified in Fig. 1.3). Furthermore, studying these connections helps identify the most influential countries, firms, or regions within the network, as well as those that are more marginal. These dimensions form the foundation of Network Science and assist in identifying and understanding complex phenomena such as delocalization, business failure, spillover effects, homophily effects, and structural properties like preferential attachment.

Advancements in network research have been instrumental in explaining these complex phenomena and enhancing our understanding of network dynamics. Researchers have made significant steps in developing novel methodologies and approaches to analyze and interpret network data. Multilevel network analysis, for example, has emerged as a powerful tool for capturing the interdependencies and interactions at different levels within complex systems (for further details, refer to the book by Lazega & Snijders, 2015). Sociologists recognized decades ago the importance of studying social systems by constructing multiple social networks using different types of ties among the same set of individuals (Wasserman & Faust, 1994; Carrington et al., 2005). In recent years, there has been an increasing focus on investigating networks with multiple types of connections, such as multiplex networks. More recently, the physics community has also contributed pioneering work on notions like networks of networks (Zhou et al., 2006) and multilayer networks

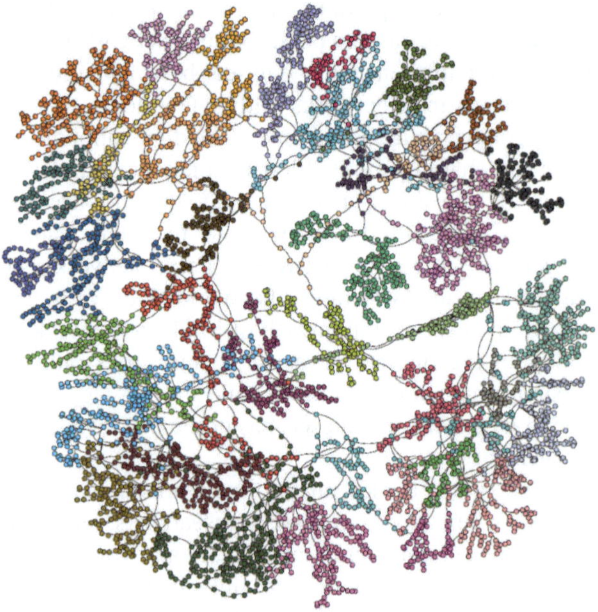

Note: author's elaboration

Fig. 1.3 Co-authorship communities. Note: author's elaboration

(Boccaletti et al., 2014; De Domenico et al., 2015). These advancements have opened up new paths for research and expanded the scope of Network Science, facilitating a deeper understanding of the underlying mechanisms driving network evolution and behavior.

Substantial progress in theoretical and epistemological approaches, driven by advancements in fields like machine learning, alongside the acknowledgment of complexity in the 2021 Nobel Prizes in Physics, and a growing emphasis on concepts such as solidarity, trust, and collaboration, have boosted the study of Network Science. Today, Network Science offers a powerful perspective for explaining numerous important phenomena. The discipline places considerable importance on endogenous causes, such as structural network configurations (e.g., popularity, homophily, reciprocity), which play a more crucial role in understanding social behavior than certain exogenous attributes like gender or age. This is because structural relationships tend to change rapidly, while individual characteristics often remain constant even as societal contexts evolve. In other words, relationships exist and dissolve within specific spatial and temporal contexts. Furthermore, relationships can be formed through various mechanisms, such as preferential attachment or homophily, and can manifest in different forms depending on the context.

It is essential to interpret these structural relationships as dynamic systems because the interactions among network members make them susceptible to frequent changes. As mentioned earlier, networks are dynamic structures that evolve rapidly. Link prediction, for example, plays a significant role in describing the evolution of complex networks. Link prediction methods aim to forecast future connections between nodes within a network. Over the years, various approaches have been developed to enhance link prediction. Among these approaches, algorithmic, matrix factorization, and machine learning, as well as hybrid approaches, have emerged as prominent strategies for predicting network links (Daud et al., 2020). Incorporating approaches like machine learning adds another dimension to understanding network dynamics. In machine learning-based link prediction, the focus is on finding the best methods and configurations to forecast future connections in a network. The goal is to minimize prediction errors as new data are introduced. By combining predictions from different approaches, the aim is to improve the accuracy and reliability of predictions. This ongoing refinement process helps enhance models over time, allowing them to better understand the evolution and dynamics of networks (Zinilli & Cerulli, 2023; Resce et al., 2022).

1.6 Network Science in Stata

Certain specific sections of the book rely on the "*nwcommands*" suite developed by Thomas Grund (2015) (https://nwcommands.wordpress.com/) for implementing the Stata software. Thomas Grund's *nwcommands* package contains a set of commands and routines for importing and manipulating networks, as well as constructing networks, computing centrality and dissimilarity measures, network visualization, and other processes. The command can be installed in a variety of ways: (i) using *findit nwcommands* and then manually install the package "*nwcommands-ado*", (ii) by writing in the command window *nwinstall, all*, or finally (iii) installing "*nwcommands-ado*" via *net install*.

Nearly 100 new Stata commands for visualizing and analyzing networks are included in the *nwcommands* suite. To make their use straightforward and practical, almost every command contains a menu with information and examples. The one caveat is that, unlike the other software we are using in this book, R and Python, the *nwcommands* suite currently supports only medium-to-small datasets. Python integration is one of the most interesting new features in Stata (from Stata 16), as it allows users to use a wide range of freely available (open-source) Python packages to process, visualize, and explore network data interactively within the Stata environment, or to directly embed Python code in Stata's *do* files. This feature, for example, enables us to employ Network Science packages that are available in Python but not yet in Stata.

1.7 Network Science in R

R is a statistical computing and graphics environment that is free to use. For Network Science research, it offers a wide range of statistical and graphical techniques. All of the offered functions may be executed via RStudio, which is an integrated development environment (IDE) for the R programming language. It runs on all platforms: Windows, Mac OS X, and Linux. The *igraph* library provides useful options for descriptive network analysis and visualization in RStudio. *install. packages("igraph")* is the command you use to install the latest version of *igraph* for RStudio. There are many other excellent packages that are not part of the *igraph*, because they solve problems in a different domain, or are designed with a different set of underlying principles. As we tackle more network data projects with RStudio, we will also use complementary packages depending on the analysis domain, such as *statnet, netsis, sna, readr, readxl, haven, Matrix, dplyr, tidytext, janeaustenr, ggraph.*

ERGMs can be generated in R using the commands *ergm* (Krivitsky et al., 2023), *ergm.count* (Krivitsky, 2015), *TERGM* and *GERGM* (Krivitsky & Handcock, 2016; Denny et al., 2017; Leifeld et al., 2018). The package *ergm.count* expands ERGMs to Poisson and geometrically distributed edge weights (discrete weights). *GERGM*, on the other hand, generalizes ERGMs when the edges are weighted with a continuous number. When analyzing longitudinal networks (e.g., panel network data), we have two options: the temporal exponential random graph model (TERGM) (Krivitsky & Handcock, 2014, 2016; Leifeld et al., 2018) and the Stochastic Actor-Oriented Model (SAOM) (Snijders 2017). TERGMs can be fitted using the package *stergm* (Krivitsky & Goodreau, 2019) or *btergm*, part of the *xergm* package suite (Leifeld et al., 2018) while SAOM can be used through the software application, *RSIENA* (Ripley et al., 2011).

1.8 Network Science in Python

The *networkX* Python package will be covered in this book. *NetworkX* is a Python library designed specifically for the creation, manipulation, and study of complex networks. If *networkX* is not currently installed on your machine, you must first install it. Use the following command to install it: *pip install networkx*. If you have an error message, you may need to use the *pip3* instead of *pip*. The *networkX* module can be used by importing it with the command: *import networkx as nx*. *NetworkX* uses Python's rich and expanding ecosystem of packages to provide additional capabilities like numerical linear algebra and graphics, such as *pandas, numpy, and matplotlib.* In this book, all codes are generated with Spyder from Anaconda Navigator, a free Python distribution that comes with all the tools you will need to run the codes. Python is mostly used to investigate networks from a statistical mechanics perspective, due to its versatility and extensive range of libraries tailored for scientific computing. Additionally, Python's libraries, such as *NetworkX*, offer robust tools for modeling and analyzing complex networks, enabling researchers to delve into statistical mechanics principles like percolation theory and Markov processes. Moreover, some integrations with other scientific libraries like *NumPy*, *SciPy*, and *Seaborn* facilitate efficient numerical computations and machine learning algorithms essential for simulating and analyzing network dynamics.

References

Barabási, A. L., & Albert, R. (1999). Emergence of scaling in random networks. *Science, 286*(5439), 509–512.

Biagini, F., Kauermann, G., & Meyer-Brandis, T. (2019). *Network science*. Springer.

Boccaletti, S., Bianconi, G., Criado, R., Del Genio, C. I., Gómez-Gardenes, J., Romance, M., et al. (2014). The structure and dynamics of multilayer networks. *Physics Reports, 544*(1), 1–122.

Borgatti, S. P., Mehra, A., Brass, D. J., & Labianca, G. (2009). Network analysis in the social sciences. *Science, 323*(5916), 892–895.

Börner, K., Sanyal, S., & Vespignani, A. (2007). Network science. *Annual Review of Information Science and Technology, 41*(1), 537–607.

Carrington, P. J., Scott, J., & Wasserman, S. (Eds.). (2005). *Models and methods in social network analysis* (Vol. 28). Cambridge University Press.

Daud, N. N., Ab Hamid, S. H., Saadoon, M., Sahran, F., & Anuar, N. B. (2020). Applications of link prediction in social networks: A review. *Journal of Network and Computer Applications, 166*, 102716.

De Domenico, M., Nicosia, V., Arenas, A., & Latora, V. (2015). Structural reducibility of multilayer networks. *Nature Communications, 6*(1), 6864.

Denny, M. J., Wilson, J. D., Cranmer, S. J., Desmarais, B. A., & Bhamidi, S. (2017). *GERGM: Estimation and fit diagnostics for generalized exponential random graph models*. R package version 0.11, 2.

Erdős, P., & Rényi, A. (1960). On the evolution of random graphs. *Publication of the Mathematical Institute of the Hungarian Academy of Sciences, 5*, 17–61.

Euler, L. (1741). Solutio Problemat is ad Geometriam Situs Pertinentis. *Commentarii Academiae Scientiarum Imperialis Petropolitanae, 8*, 128–140.

Filippetti, A., & Zinilli, A. (2023). The innovation networks of city-regions in Europe: Exclusive clubs or inclusive hubs? *Papers in Regional Science, 102*(6), 1169–1193.

Freeman, L. (2004). *The development of social network analysis: A study in the sociology of science*. Empirical Press.

Freeman, L. C., White, D. R., & Romney, A. K. (1989). *Research methods in social network analysis*. Transaction Books.

Granovetter, M. S. (1973). The strength of weak ties. *American Journal of Sociology, 78*(6), 1360–1380.

Grund, T. (2015). *Social network analysis using Stata*. United Kingdom Stata users' Group meetings 2015, 21, Stata Users Group.

Huisman, M. (2009). Imputation of missing network data: Some simple procedures. *Journal of Social Structure, 10*(1), 1–29.

Karinthy, F. (1929). *Chain-links. Everything is different*, 21–26.

Krivitsky, P. (2015). *Ergm. count: Fit, simulate and diagnose exponential-family models for networks with count edges*. The Statnet Project.

Krivitsky, P. N., & Goodreau, S. M. (2019, June). *STERGM-separable temporal ERGMs for modeling discrete relational dynamics with statnet*. Tech. Rep. Retrieved from http://statnet.csde.washington.edu/workshops/SUNBELT

Krivitsky, P. N., & Handcock, M. S. (2014). A separable model for dynamic networks. Journal of the Royal Statistical Society. Series B. *Statistical Methodology, 76*(1), 29.

Krivitsky, P. N., & Handcock, M. S. (2016). *tergm: Fit, simulate and diagnose models for network evolution based on exponential-family random graph models*. The Statnet Project (http://www.statnet.org). R package version, 3(0).

Krivitsky, P. N., Hunter, D. R., Morris, M., & Klumb, C. (2023). Ergm 4: New features for analyzing exponential-family random graph models. *Journal of Statistical Software, 105*, 1–44.

Lazega, E., & Snijders, T. A. B. (dir.) (2015). *Multilevel network analysis for the social sciences. Theory, methods and applications*, Springer, series: « Methodos », 2015, 375 p., ISBN: 978-3-319-24518-8.

Leifeld, P., Cranmer, S. J., & Desmarais, B. A. (2018). Temporal exponential random graph models with BTERGM: Estimation and bootstrap confidence intervals. *Journal of Statistical Software, 83*(6).

Milgram, S. (1967a). Six degrees of separation. *Psychology Today,* 2, 60–64.

Milgram, S. (1967b). The small world problem. *Psychology Today, 2*(1), 60–67.

Moreno, J. L. (1934). *Who shall survive?: A new approach to the problem of human interrelations*. Nervous and Mental Disease Publishing Co.. https://doi.org/10.1037/10648-000

Moreno, J. L. (1947). Sociometry and the social psychology of G.H. Mead. *Sociometry, 10*, 350–353.

Newman, M. E. J. (2003). The structure and function of complex networks. *SIAM Review, 45*, 167–256. https://doi.org/10.1137/S003614450342480

Resce, G., Zinilli, A., & Cerulli, G. (2022). Machine learning prediction of academic collaboration networks. *Scientific Reports, 12*, 21993. https://doi.org/10.1038/s41598-022-26531-1

Ripley, R. M., Snijders, T. A., Boda, Z., Vörös, A., & Preciado, P. (2011). *Manual for RSIENA* (Vol. 1, p. 2011). University of Oxford, Department of Statistics, Nuffield College.

Snijders, T. A. (2017). Stochastic actor-oriented models for network dynamics. *Annual Review of Statistics and Its Application, 4*, 343–363.

Strauss, C. L. (1974). *Structural anthropology* (Vol. 1, p. 571). Persona & Derecho.

Wasserman, S., & Faust, K. (1994). *Social network analysis: Methods and applications*. Cambridge University Press.

Watts, D. (2003). *Six degrees: The science of a connected age*. Norton.

Watts, D. J., & Strogatz, S. H. (1998). Collective dynamics of 'small-world' networks. *Nature, 393*(6684), 440–442.

Zhou, C., Zemanová, L., Zamora, G., Hilgetag, C. C., & Kurths, J. (2006). Hierarchical organization unveiled by functional connectivity in complex brain networks. *Physical Review Letters, 97*(23), 238103.

Zinilli, A. (2016). Competitive project funding and dynamic complex networks: Evidence from projects of National Interest (PRIN). *Scientometrics, 108*(2), 633–652.

Zinilli, A., & Cerulli, G. (2023). Link prediction and feature relevance in knowledge networks: A machine learning approach. *PLoS One, 18*(11), e0290018.

Chapter 2
Network Science: Concepts and Definitions

2.1 Graphs: Notation and Simple Matrix Algebra

The word "graph" has at least two different meanings in mathematics. In algebra and calculus, a graph of a function is a continuous line chart or surface plot, while in graph theory, a graph is a set of discrete objects connected by lines or arcs. In this book, we will always use the latter definition. A graph, or network, consists of objects called vertices or nodes, connected by arcs known as edges, links, or ties. These edges indicate the presence of a connection between the vertices. Usually, the terms network, node, and link refer to real systems, while the terms graph, vertex, and edge are used when describing the mathematical representation of these networks. In scientific literature, the terms network and graph are often used interchangeably. We will adopt the same convention throughout this book.

The main terminology used in the literature of graphs or network theory is presented here. Much of modern mathematics deals with the interactions between objects, or more generally, with how elements within the same class relate to each other. Networks offer a clear and intuitive way to represent these relationships.

In the following graph G (Fig. 2.1), the vertex set is $V = \{1,2,\ldots,7\}$ and the edge set is $E = \{(1,2), (1,4), (1,5), (2,5), (2,6), (3,5), (3,6), (3,7), (4,6), (4,7)\}$.

Definition Formally, a graph G is defined as a pair $G = (V, E)$, where V is a finite (non-empty) set of vertices (or nodes) and E is the set of edges, where the items of E are unordered pairs $\{i, j\}$ of distinct vertices $i, j \in V$. Each edge joins two vertices i and j.

Vertices represent the components in a graph. The size of the network is commonly referred to as N. Usually, we label the vertices with $i = 1, 2, \ldots, n$. Interactions between the nodes are represented by the edges, which connect the vertices.

The following are some examples of special graphs:

A. Zinilli, *Elements of Network Science*, Statistics and Computing,
https://doi.org/10.1007/978-3-031-84712-7_2

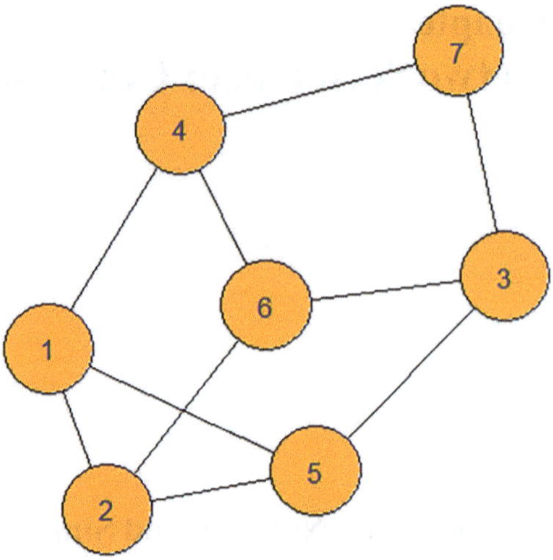

Fig. 2.1 A graph with vertices = 7 and edges = 10

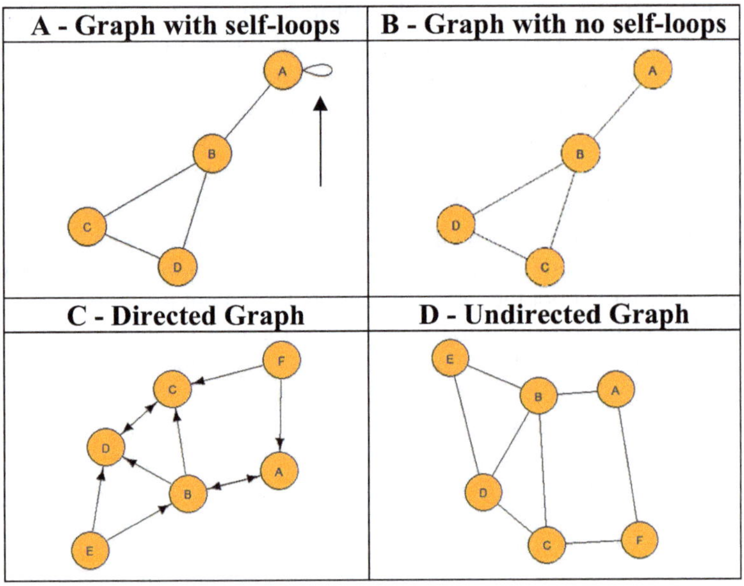

Fig. 2.2 An example of special graphs. (**a**) Graph with self-loops. (**b**) Graph with no self-loops. (**c**) Directed Graph. (**d**) Undirected Graph

Graph with Self-Loops This indicates that one can return to the same vertex via an edge without leaving it (Fig. 2.2a). This occurs when the set of edges E fulfills the following constraint:

$$\exists i \in V, (i,i) \in V \tag{2.1}$$

Graph with No Self-Loops This means that it is impossible to return to the same vertex in a single transition (Fig. 2.2b). The set of edges E is irreflexive, namely:

$$\forall i \in V, (i,i) \notin E \tag{2.2}$$

Directed Graph Or even oriented network: It is a network where the edges are indicated by the ordered pair $(i, j) \in E$. However, if the edge (i, j) is present, the reverse edge (j, i) is not required. Usually, the ordered structure of the edges is typically represented by an arrow, denoting the direction of an edge (Fig. 2.2c).

Undirected Graph Also called non-oriented network, it is a network where an edge is an unordered pair $\{i, j\} \in E$ indicating that i and j are directly connected. If i is directly connected to j when G is undirected, then j is also directly connected to i (Fig. 2.2d).

Complete Graph A complete graph is one in which every pair of vertices $\{i, j\}$ is linked. When a graph is undirected, a complete graph is equal to $|E| = N(N\text{-}1)/2$, N represents the total number of vertices in the graph (Fig. 2.3a). When the graph is directed, a complete graph is equal to $|E| = N(N\text{-}1)$.

A complete graph is a simple graph in which every pair of distinct vertices is connected by exactly one edge, without self-loops or multiple edges.

Null Graph A null graph is a graph that has no edges (Fig. 2.3b). In this specific context, even if the edge set is empty, the vertex set cannot be empty.

Isomorphism A graph is essentially the expression of a symmetrical binary relation (defined by the set E) on V, independently from the nature of the elements of V. An isomorphism between two graphs $G = (V, E)$, and $G' = (V', E')$, is a bijection function $f: V \rightarrow V'$ between sets of vertices such that $\{i, j\} \in E$ if and only if $\{f(i)\ f(j)\} \in E'$. As a result, an isomorphism merely relabels the vertices, rendering the networks identical except for the relabeling. Isomorphism, therefore, refers to the

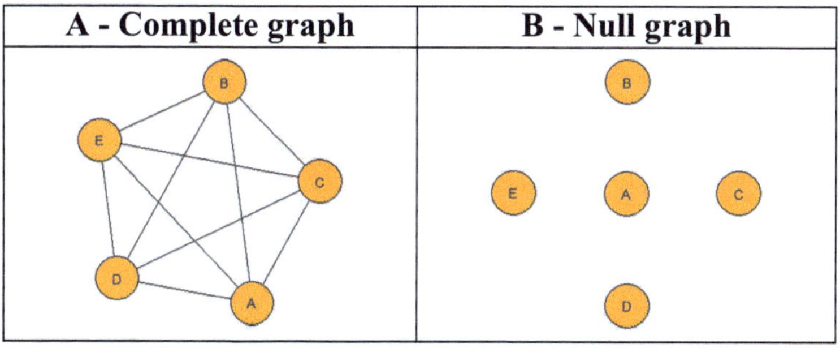

Fig. 2.3 An example of special graphs. (**a**) Complete graph. (**b**) Null graph

notion that two networks or graphs share the same structure, even if their vertex labels differ (Fig. 2.4).

In the previous example (Fig. 2.4), the number of vertices and edges in both graphs is the same, precisely 4. Both graphs have the same degree sequence (degree 2). Finally, the vertices in both graphs form a 4-length cycle. The graphs in the example are isomorphic, which implies they are equivalent graphs in various forms.

Tree Graph A tree graph is a connected, acyclic graph where each node (except the root) has exactly one parent. A rooted tree has a designated root node from which all other nodes descend. In any tree, the number of vertices is equal to the number of edges plus one, i.e., $V = E+1$. An example of a tree graph is shown in Fig. 2.5.

Multigraph A multigraph is a graph that may include multiple edges between the same pair of vertices and may also include loops (edges that connect a vertex to itself).

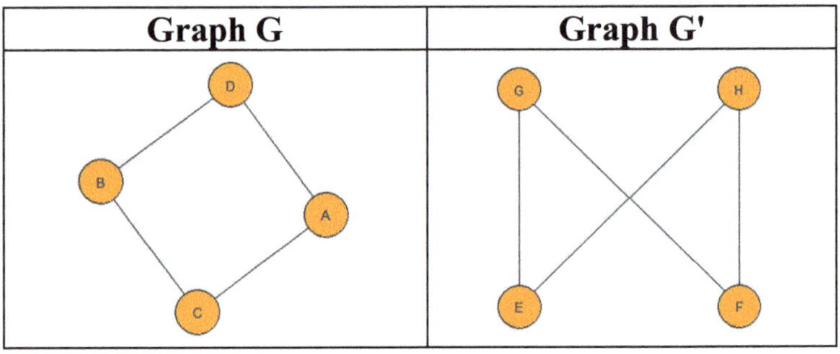

Fig. 2.4 An example of isomorphism. (**a**) Graph G. (**b**) Graph G′

Fig. 2.5 Tree graph with V = 10 and E = 9

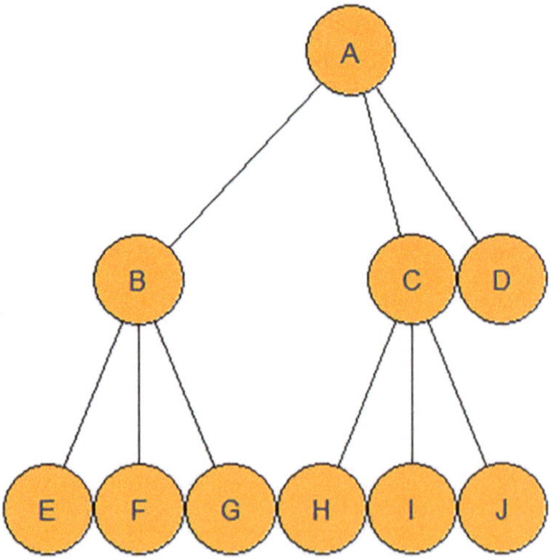

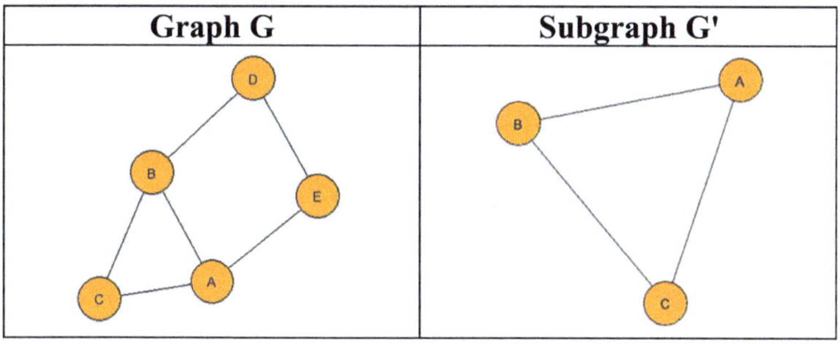

Fig. 2.6 An example of subgraph. (**a**) Graph G. (**b**) Subgraph G′

Subgraph A graph $G' = (V', E')$, is said to be a subgraph of the graph $G = (V, E)$ if all the vertices in V' belong to V and all the edges in E' belong to E, that is, $V' \subseteq V$ and $E' \subseteq E$. The graphs shown in Fig. 2.6, G and G', represent, respectively, the original graph and one of its subgraphs.

Subgraph (Dyads) A dyad is a subgraph that represents a pair of actors and the possible edge between them. In an undirected relation represented as a graph, an unordered pair of vertices can only be in one of two states: either they are adjacent (connected by an edge) or they are not (not connected by an edge) (Fig. 2.7).

Subgraph (Triads)
A triad is a subgraph made up of three vertices and the possible connections between them. A triad in a graph can have four states, depending on whether zero, one, two, or three edges connect the three vertices in the triad (Fig. 2.8).

Graphs can also be classified based on the number of edges they include.

Unweighted Graph An unweighted graph G is one in which $E = 1$ for every edge connecting vertex i and j; $E = 0$ if no edge exists between vertex i and j (Fig. 2.9a).

Weighted Graph A weighted graph G is defined as $G = (V, E, W)$, where V and E are the sets of vertices and edges, respectively, and W represents the edge weights. Usually, the edge weights are nonnegative numbers $w > 0$, and the weighted graphs may be either directed or undirected (Fig. 2.9b).

Bipartite graph, also called a two-mode network, is a graph (N, g) in which the set of vertices N may be partitioned into two sets (or modes) A and B, such that if an edge (i,j) belongs to g, then one vertex is from set A and the other from set B (Opsahl, 2013). Figure 2.10 displays a bipartite graph.

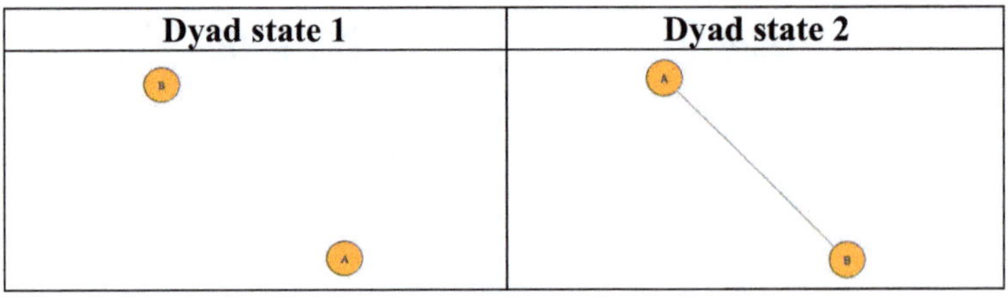

Fig. 2.7 Dyad states. (**a**) Dyad state 1. (**b**) Dyad state 2

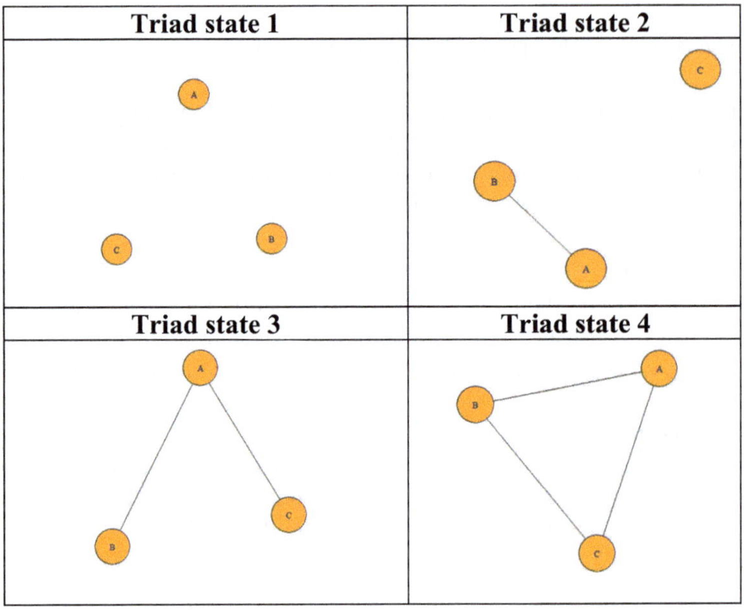

Fig. 2.8 Triad states. (**a**) Triad state 1. (**b**) Triad state 2. (**c**) Triad state 3. (**d**) Triad state 4

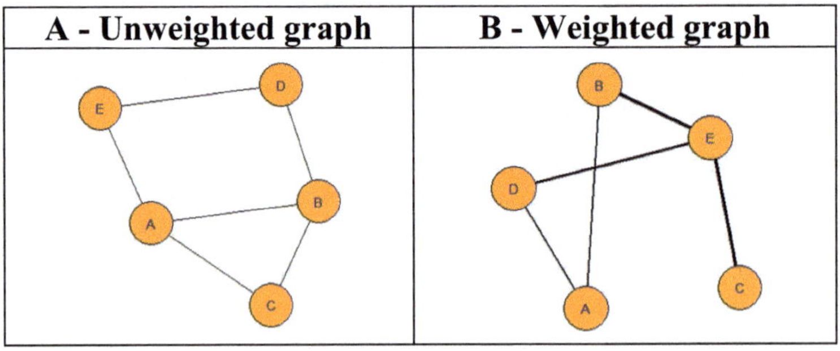

Fig. 2.9 An example of an unweighted and a weighted graph. (**a**) Unweighted graph. (**b**) Weighted graph

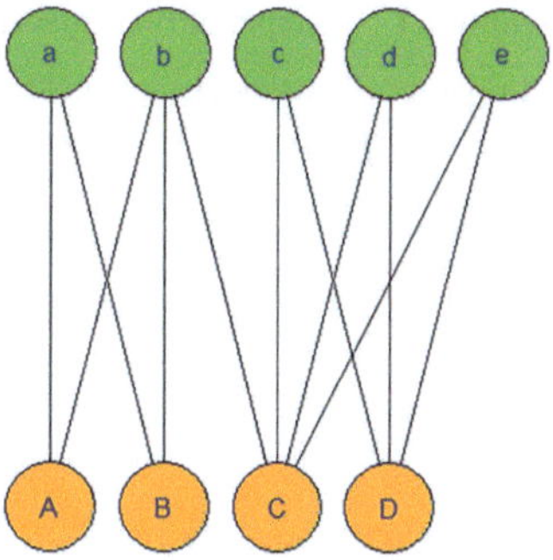

Fig. 2.10 An example of bipartite graph

2.2 Handling Network Data

The adjacency matrix is an $n \times n$ square matrix, where n denotes the number of vertices in the graph. If there is a connection (i.e., an edge) between two vertices, a 1 will occur in the matrix; otherwise, a 0 appears. In the case of undirected graphs, the adjacency matrix is symmetrical, and the edges incident on the generic vertex i may be measured by summing the elements of the *i-th* row, while the edges incident on the generic vertex j can be measured by summing the elements of the *j-th* column (Table 2.1a).

In the case of directed graphs, this matrix loses its symmetry (Table 2.1b) (Borgatti et al., 2022). Summing the components of a generic *i-th* row yields the number of edges departing from a generic vertex i. Summing the elements of a generic *j-th* column yields the number of edges entering a generic vertex j.

The properties of an adjacency matrix for a non-directed graph can be summarized as follows:

$$\begin{aligned} &1)\quad G_{ii} = 0 \\ &2)\quad G_{ij} = G_{ji} \\ &3)\quad E = \frac{1}{2} \sum_{V}^{i,j=1} G_{ij} \end{aligned} \tag{2.3}$$

The properties of an adjacency matrix for a directed graph can be summarized as follows:

$$\begin{aligned} &1)\quad G_{ii} \neq 0 \\ &2)\quad G_{ij} \neq G_{ji} \\ &3)\quad E = \sum_{V}^{i,j=1} G_{ij} \end{aligned} \tag{2.4}$$

The incidence matrix is an $n \times m$ matrix, where the number of rows corresponds to the number of vertices and the number of columns to the number of edges. If the

Table 2.1 Example of two adjacency matrices

(a) Non-oriented adjacency matrix	(b) Oriented adjacency matrix
$G = \begin{bmatrix} 0 & 2 & 1 & 0 \\ 2 & 0 & 0 & 1 \\ 1 & 0 & 0 & 3 \\ 0 & 1 & 3 & 0 \end{bmatrix}$	$G = \begin{bmatrix} 0 & 2 & 0 & 3 \\ 0 & 0 & 0 & 1 \\ 0 & 0 & 0 & 1 \\ 1 & 2 & 0 & 0 \end{bmatrix}$

Table 2.2 Example of incidence matrices

(a) Non-oriented incidence matrix	(b) Oriented incidence matrix
$G = \begin{bmatrix} 1 & 1 & 0 & 0 & 0 & 0 \\ 1 & 0 & 1 & 1 & 0 & 0 \\ 0 & 1 & 1 & 0 & 1 & 1 \\ 0 & 0 & 0 & 1 & 1 & 0 \end{bmatrix}$	$G = \begin{bmatrix} 1 & 1 & 0 & 0 & 0 & 0 \\ -1 & 0 & -1 & 1 & 0 & 0 \\ 0 & -1 & 1 & 0 & 1 & 1 \\ 0 & 0 & 0 & -1 & -1 & 0 \end{bmatrix}$

Table 2.3 Example of an adjacency list

Adjacency list
1→2→3
2→3
3→1→4
4→2

represented graph is undirected (Table 2.2a), the matrix elements will be equal to 1 in the case of an incidence relationship between the vertex and the edge, and 0 in the case where the edge has no implication on the vertex. In the case of directed graphs (Table 2.2b), the notation differs because 1 is used when the edge affects the inbound vertex and -1 when the edge leaves the vertex. When there is no connection between the vertex and the edge, the value 0 is used.

The last method of representation is the adjacency list, which lists each vertex along with its adjacent vertices. This method is especially useful for representing sparse graphs, which are common in real-world networks where the number of edges is small compared to the number of vertices. In such cases, an adjacency matrix would be large and mostly filled with zeros, since it explicitly represents both existing and non-existing edges. This leads to inefficient use of memory and computational resources. In contrast, adjacency lists store only the existing edges, making them more efficient (Table 2.3).

An edge list (Table 2.4) is a variation of graph representation. It consists of a simple two-column list, where each row represents an edge by specifying the pair of connected vertices.

Table 2.4 Example of an edge list

Edge list
1→2
1→3
1→4
2→3
2→4
3→1
3→4
4→2

Using both adjacency and edge list implementations can significantly reduce computation time. All of the software we will be using in this book supports encoding in both matrix and list formats, allowing for easy switching between them with just a few lines of code.

2.3 Import and Generate Network Data in Stata, R, and Python

This section explains how to import network data into the three software platforms used throughout this book: Stata, R, and Python. The first step involves importing the dataset using specific libraries and packages available in each environment, which will be introduced at the beginning of the relevant sections. This section is organized into three parts: (1) data import from various sources, (2) data processing and network creation, and (3) network visualization, with an emphasis on key components.

In Stata, we primarily rely on the macro package *nwcommands*, supplemented by the *datanet*, *netsis*, and *netsummarize* methods (Zinilli & Cerulli, 2017). We will also illustrate how to create adjacency matrices using *Mata*, a matrix programming language that augments Stata's functionality.

R offers two popular packages for network manipulation: *statnet* and *igraph* (Kolaczyk & Csárdi, 2014). Python, on the other hand, boasts two major modules—*pandas* and *networkX* —that facilitate manipulation, visualization, and analysis of network data (McKinney, 2012; Al-Taie & Kadry, 2017) .

All the packages for Network Science in Stata, R, and Python comprise graph object classes, generators for creating full graphs, routines for reading in real data, algorithms for analyzing the resulting networks, and basic graphical tools.

2.3.1 *Stata Code and Output*

From a MATA Adjacency Matrix

```
// Search for nwcommands
. findit nwcommands

// Having information on the nwcommands routines
. help nwcommands

// Creating matrix A in Mata
. mata A=(0,10,1 \5,0,0 \0,2,0)

// Observing the new matrix A
. mata A

        1    2    3
    +----------------+
  1 |   0   10    1  |
  2 |   5    0    0  |
  3 |   0    2    0  |
    +----------------+

// Export matrix A from MATA to Stata. Stub option assigns the name to variable identifying the membership
. nwtostata, mat (A) stub (n)

// Declare data to be network data
. nwset , mat(A)

// Details on the new matrix
. nwnodelab, nodeid(3) detail

// Declare data to be network data and with name() we assign a name to STATA network
. nwset, mat(A) name(netA)
```

From a Stata Adjacency Matrix

```
// In Stata, we create a directed adjacency matrix. label identifies the name of vertices

n1      n2      n3      label
0       10      1       x
5       0       0       y
0       2       0       z

// Plot the network just created. We use the option arcsyle() for having curved edges
. nwplot,label(matrix) arcstyle(curved)

// Edge size by weight
. nwplot netA, edgesize(netA)
```

From a Stata Edge List

```
// Clear all networks and variables from memory
. nwset, nwclear

// Create two variables (ego and alter) that will be used to identify the relationships: (1, 2), (2, 3), (2, 4), (3, 4)

ego     alter
 1        2
 2        3
 2        4
 3        4
 5        5

// Declare data in edge list format
. nwset ego alter, edgelist

// Plot the network with labels
. nwplot, lab
```

List of all Networks in Memory

```
// List networks in memory
. nwds

// Display more details about existing networks
. nwset, detail

// Show the current network. The current network is just the most recently set network through the command nwset
. nwcurrent
```

Converting an Edge List to an Adjacency Matrix

```
// The example below shows how to make an edge list in Stata and convert it to an adjacency matrix and vice versa.

// Edge list creation
. nwset, nwclear
. gen ego=.
. set obs 1
. replace ego = 1 in 1
.
.

// Edge list converted to adjacency matrix
. nwfromedge ego alter, name(mynet)

// To create an edge list from the newly formed adjancecy matrix
. nwtoedge mynet

ego   alter              _nodelab  _nodevar  _nodeid  net1  net2  net3  net4  net5
1     2                  1         net1      1        0     1     0     0     0
2     3     ———————>     2         net2      2        0     0     1     1     0
2     4                  3         net3      3        0     0     0     1     0
3     4                  4         net4      4        0     0     0     0     0
5     5                  5         net5      5        0     0     0     0     1
```

2.3.2 R Code and Output

Edge List Creation

```
# Install and upload the library
> install.packages("igraph")
> library(igraph)

# We set the seed (random number), which is useful for creating random objects that can be reproduced
> set.seed(99)

# Create a dataframe "list "with two columns
> list <- data.frame(source=c("A", "A", "A", "A", "B", "F", "B"), target=c("B", "B", "C", "D", "E", "A", "F"))

# We build the graph object that we call "network"
> network <- graph_from_data_frame(d=list, directed=F)

# Plot the network
> plot(network)
```

Adjacency Matrix Creation #1

```
# Install and upload the library
> install.packages("igraph")
> library(igraph)

# We set the seed (random number), which is useful for creating random objects that can be reproduced
> set.seed(99)

# Create a 5x5 matrix with values from 0 to 1
> matrix <- matrix(sample(0:1, 25, replace=TRUE), nrow=5)

# The first five alphabetic letters are used to label the columns and rows.
> colnames(matrix) = rownames(matrix) = LETTERS[1:5]

# We build the graph object that we call "network"
> network <- graph_from_adjacency_matrix(data)

# Plot the network
> plot(network)
```

Adjacency Matrix Creation #2

```
# Install and upload the library
> library(statnet)
> library(sna)
> library(readr)
> library(readxl)

# We set the seed (random number)
> set.seed(999)

# We set the number of nodes
> num_nodes <- 26

# The runif() function is used to generate n uniform random numbers. nrow and ncol must be the same.
> matrix<-matrix(round(runif(num_nodes*num_nodes)), nrow = num_nodes, ncol = num_nodes)

# We replace the diagonal elements of the matrix with 0 for avoiding self -loops
> diag(matrix) <- 0

# We build the graph object that we call "net". We specify whether the network is directed or undirected. Option matrix.type is the typology of input
> net <- as.network(x = matrix, directed = FALSE, loops = FALSE, matrix.type = "adjacency")

# We give each of the 26 nodes in my network a name. The labels in this case are the ISO codes of several European countries.
> network.vertex.names(net) <- c("AT", "BE", "CH", "CY", "DE", "DK", "EE", "EL", "ES", "FI", "FR", "HU", "IE", "IL", "IT", "LT", "LU", "LV", "MT", "NL", "NO", "PL", "PT", "RO", "SE", "SK")
```

Import Data from External Sources

```
# haven package allows R to read and write a variety of data formats used by other statistical tools. readr package provides a fast way
to read data formats like 'txt', 'csv', 'tsv', and 'fwf'
> install.packages("haven")
> install.packages("readr ")

# We download the libraries
> library(haven)
> library("readr")

# setwd() function sets the working directory
> setwd("C:\\mypath")

# We import the dataset we want to use
> dataset<- read.csv("C:\\mypath\\dataset-name.txt")

# Splitting the dataset by the "year" variable
> split<-split(dataset, dataset$year)

# We can see that several datasets have been created
> str(split)

# Now we can create four different datasets
> y2011<-split$'2011'
> y2012<-split$'2012'
> y2013<-split$'2013'
> y2014<-split$'2014'
```

2.3.3 Python Code and Output

Graph Creation #1

```
# Loading Python Libraries. 'as' is a more practical way to use the libraries. It can be used to call the specific library, related
functions and data types
>>> import pandas as pd
>>> import numpy as np
>>> import networkx as nx
>>> import matplotlib.pyplot as plt

# Create an empty graph G (undirected graph). We can write G = nx.DiGraph() to create an empty direct graph
>>> G = nx.Graph()

# Add one node
>>> G.add_node(1)

# Add nodes from a list
>>> G.add_nodes_from([2, 3])

# We add one edge between node 1 and 2
>>> G.add_edge(1, 2)

# Add an edge between node 2 and 3
>>> e = (2, 3)
>>> G.add_edge(*e)

# Plot the network with label option
>>> nx.draw(G, pos=nx.circular_layout(G), node_color='r', edge_color='b', with_labels=True)
```

Graph Creation #2

```
# Loading Python Libraries
>>> import networkx as nx
>>> import matplotlib.pyplot as plt
>>> %matplotlib inline
>>> import warnings; warnings.simplefilter('ignore')

# We build the symmetric network. The 'weight' option allows you to add the relationship's intensity to the network; each edge has a weight. We use nx.DiGraph() for directed networks
>>> MyGraph = nx.Graph()
>>> MyGraph.add_edge('Antonio', 'Valentina')
>>> MyGraph.add_edge('Antonio', 'Claudia')
>>> MyGraph.add_edge('Claudia','Valentina')

# Some network info are visualized
>>> print(nx.info(MyGraph))

# We use the draw_networkx() function to visualize the network and save it to the folder.nx.draw_networkx(MyGraph, font_size=7)
>>> fig.savefig('myGraph.png', dpi=100)
```

Edge List Creation

```
# Loading Python libraries
>>> import pandas as pd
>>> import numpy as np
>>> import networkx as nx
>>> import matplotlib.pyplot as plt

# Build a dataframe with 4 connections
>>> df = pd.DataFrame({ 'from':['A', 'B', 'C','A'], 'to':['D', 'A', 'E','C']})

# Build the graph
>>> G=nx.from_pandas_edgelist(df, 'from', 'to')

# Plot it
>>> nx.draw(G, with_labels=True)
>>> plt.show()
```

Adjacency Matrix Creation

```
# Loading Python Libraries
>>> import pandas as pd
>>> import numpy as np
>>> import networkx as nx
>>> import matplotlib.pyplot as plt

# Create a graph in adjacency format
>>> graph = {'A': set(['B', 'C']),
        'B': set(['A', 'D', 'E']),
        'C': set(['A', 'F']),
        'D': set(['B']),
        'E': set(['B', 'F']),
        'F': set(['C', 'E'])
       }

# Save the generated figure in .png format. We can also save it as EPS, JPEG and other formats
>>> plt.savefig("graph.png")
```

From Edge List to Adjacency Matrix

```
# We create a list of 4 items
>>> E = [[0, 0], [0, 1], [1, 0], [1, 1]]

# We indicate the size of each item in E
>>> size = len(set([n for e in E for n in e]))

# Make an empty adjacency list
>>> adjacency = [[0]*size for _ in range(size)]

# Populate the list for each edge
>>> for sink, source in E:
    adjacency[sink][source] = 1

# Adjacency matrix visualization
>>> print(adjacency)
```

2.4 Bipartite Network and Projection to One-Mode

Bipartite networks (or bipartite graphs) are an important class of complex networks. Many social and economic systems can be represented as bipartite networks, such as co-patent and co-authorship networks, trade networks, and migration networks. A bipartite network is a graph containing two unique sets of vertices that may be separated into two independent sets (partitions). It is frequently represented as $G = (V1 + V2, E)$, where *V1* and *V2* are the two independent sets (two different objects). Only vertices from separate partitions can form edges; vertices in the same partition are not allowed to interact with each other. In real-life contexts, bipartite networks are rarely investigated in their original form. As a result, these networks are frequently analyzed as one-mode networks (with only one type of vertex); projection is the term used to describe this operation. Projection is accomplished by choosing one of the vertex sets and connecting two vertices from that set if they are linked to the same vertex of the other kind. An example of projection can be seen in Fig. 2.11.

In the example in Fig. 2.11, nodes *A* and *B* are connected because they share at least one green vertex. Weights can be assigned to edges in projected one-mode networks.

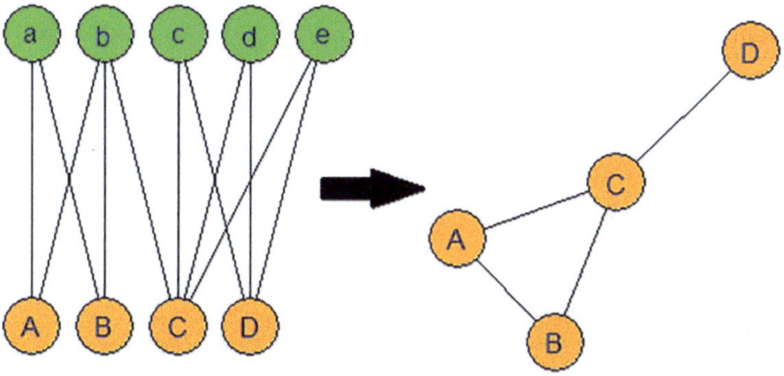

Fig. 2.11 An example of projection

It is feasible to generate a weighted one-mode network in all three software (Stata, R, and Python) by defining the weights as the number of co-occurrences (e.g., the number of patents that two authors had collaborated on). The creation of a one-mode network from a two-mode or bipartite network would aid in discovering previously unknown social interactions between vertices in the same set (Zinilli & Cerulli, 2017).

2.4.1 Stata Code and Output

One-Mode Projection with Nwcommands

```
// Declares the data as two-mode
. nw2fromedge level1 level2 [tievalue], name(mynet)

// Make one-mode projection to either level 1 or level 2
. nw2fromedge level1 level2, project(2)
```

One-Mode Projection with Datanet Command

```
// In this case the variable level1 must be string
. datanet level1 level2, save(mynetwork)
```

2.4.2 R Code and Output

```
# Loading R Libraries
 > library(igraph)
 > library(readxl)
 > library(Matrix)
 > library(dplyr)

# We define memory limit (only for large datasets)
> memory.limit(9000)

# In bipartite format, we create a dataframe with the two variables identifying the vertices. A network can consist of different
'classes' of vertices. For example, a bipartite network might consist of organizations (the first column) and common projects in
which they are members (the second column).
> mydata <- data.frame(dataset)

# Table function uses the cross-classifying factors to build a contingency table of the counts at each combination of factor levels
> m <- table(mydata)

# As.matrix transforms the argument previously created into a matrix
> Matrix <- as.matrix(m)

# To get the one-mode representation of ties between rows, multiply the matrix by its transpose. Mrow will now represent the one-
mode matrix formed by the row entities (the organizations in our case)
> Mrow <- tcrossprod(A)) # other option
> Mrow <- Matrix %*% Matrix
>

# To get the one-mode representation of ties between columns, multiply the matrix by its transpose. Mcol will now represent the one-
mode matrix formed by the column entities (the projects in our case)
> Mcol <- tcrossprod(t(A)) # Other option
> Mcol <- t(Matrix) %*% Matrix

# We make the new matrix's diagonal equal to 0
> diag(Mrow) <- 0
```

2.4.3 Python Code and Output

```
# There is no specific bipartite graph class in NetworkX, however bipartite graphs may be represented using the Graph() or
DiGraph() classes.

# Loading Python library
>>> from networkx.algorithms import bipartite
>>> import networkx as nx
>>> import matplotlib.pyplot as plt

# Create an empty graph with no vertices and no edges
>>> A = nx.Graph()

# We add the edge list of the two entities (letters and numbers)
>>> A.add_edges_from([("a", 1), ("b", 1), ("a", 2), ("c", 2)])

# Returns the projection of A
>>> G = bipartite.projected_graph(A, ["a", "b", "c"])

# Plot the graph in one-mode format
>>> nx.draw(G, with_labels=True)
>>> plt.show()
```

2.5 Network Visualization

Network visualization involves representing the relationships between vertices. Among scientific methods, network visualization is notable for its strong visual and often artistic dimension. This section will aid in the process of visualizing networks in each of the three software platforms. The main options for improving the quality of representation will also be illustrated. In Stata, a network is represented by the macro command *nwcommands*. The primary tool in R for visualizing and displaying a network is the package *igraph*. The *igraph* package analyzes and visualizes large networks effectively. Finally, *matplotlib* will be used for Python visualization.

The settings shown in the example boxes below are just a few of the many options that we can use in graph visualization.

2.5.1 Stata Code and Output

```
// Clear all networks and variables from memory
. clear all
. nwset, nwclear

// Loading the florentine data
. webnwuse florentine, nwclear

// Plot the network. The "wealth" variable associated with the network is used to color the nodes. The legend position and the size in
the plot window is determined by the option legendopt
. nwplot flomarriage, color(wealth) legendopt(on pos(6) rows (3) cols(2) size(vsmall))

// Plot the network as a sociomatrix, sorting the "wealth" variable label
. nwplotmatrix flomarriage, sortby(wealth) label(wealth)
```

2.5.2 R Code and Output

```
# Loading igraph library
> library(igraph)

# Create data and the graph adjacency matrix
> data <- matrix(sample(0:1, 100, replace=TRUE, prob=c(0.8,0.2)), nrow=25, ncol=25)
> network <- graph_from_adjacency_matrix(data, mode='undirected', diag=F )

# We can use multiple layouts and parameters at the vertex and edge level when plotting
> par(mfrow=c(2,2), mar=c(1,1,1,1))
> plot(network, layout=layout.sphere, main="Sphere", vertex.color="red", vertex.size=25, edge.color="black", edge.size=1,
vertex.label.family="Times", edge.width=1 , vertex.color = rgb(1,0.4,0))
> plot(network, layout=layout.circle, main="Circle", vertex.color="orange", vertex.size=25, edge.color="black", edge.size=1
,vertex.label.family="Times", edge.width=1 , vertex.color = rgb(1,0.4,0))
> plot(network, layout=layout.random, main="Random", vertex.color="green", vertex.size=25, edge.color="black",
edge.size=1, vertex.label.family="Times", edge.width=1 , vertex.color = rgb(1,0.4,0))
> plot(network, layout=layout.fruchterman.reingold, main="Fruchterman.reingold", vertex.color="violet", vertex.size=25,
edge.color="black", edge.size=1, vertex.label.family="Times", edge.width=1 , vertex.color = rgb(1,0.4,0))

Output
```

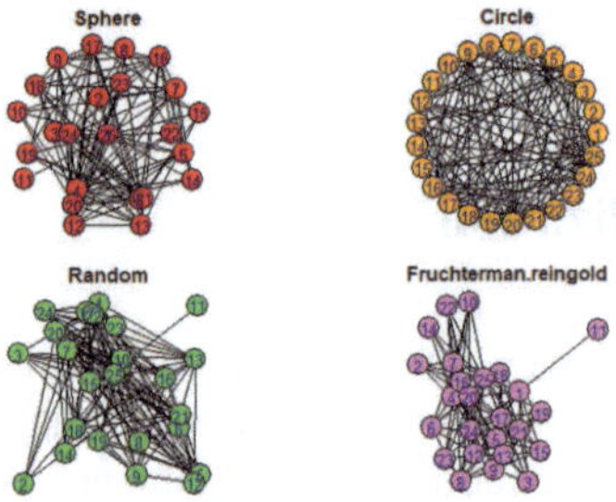

Bipartite Graph Visualization

```
# Loading igraph library
> library(igraph)

# Create data
> g <- graph(c(1,2,1,3))
> par(mar=c(0,0,0,0))
> A <- matrix(c(2,4,0,0,1,2,1,0,0,0,4,5,0,0,1,2,0,0,6,2), byro w = T, nrow = 5, ncol = 4)
> rownames(A) <- letters[c(1:nrow(A))]
> colnames(A) <- LETTERS[c(1:ncol(A))]
> print(A)

# Create graph from an incidence matrix
> g <- graph.incidence(A, weighted = TRUE)

# Set "type" attribute by vertex
> V(g)$type

# Set "name" attribute by vertex
> V(g)$name
#Set the color of the nodes by "type" level
> V(g)$color <- V(g)$type
> V(g)$color=gsub("FALSE","green",V(g)$color)
> V(g)$color=gsub("TRUE","orange",V(g)$color)

# Plot the bipartite network specifying options at vertex and edge level
> plot(g, vertex.size=40, edge.color="black", edge.size=1, edge.arrow.size=0, vertex.label.family="Times", edge.width=0 ,
layout=layout_as_bipartite)
```

2.5.3 Python Code and Output

```
# Loading Python libraries
>>> import pandas as pd
>>> import numpy as np
>>> import networkx as nx
>>> import matplotlib.pyplot as plt

# Create data
>>> MyGraph = nx.Graph()
>>> MyGraph.add_edge('Antonio', 'Valentina')
>>> MyGraph.add_edge('Antonio', 'Claudia')
>>> MyGraph.add_edge('Claudia', 'Valentina')
>>> print(nx.info(MyGraph))

# We can use multiple layouts (nx.draw_circular, nx.draw_planar, nx.draw_random, nx.draw_spectral, nx.draw_spring,
nx.draw_shell)
>>> nx.draw_planar(MyGraph, with_labels = True, font_size=10, node_color='red', edge_color='blue')
```

2.6 Text Networks

Although network analysis in social sciences is most frequently used to represent connections between people, organizations, regions, and countries, it may also be used to describe relationships between words. Text networks are commonly used in graph-based natural language processing, such as keyword extraction. A corpus of texts, for example, may be shown as a network with each node representing a document and the thickness or strength of the edges between them indicating similarities between the words used in any two documents. Words can also be nodes in a text network, with the edges connecting them reflecting how frequently they appear in documents. A text network, or word co-occurrence network, is a graph of word interactions representing the co-occurrence of words in a specific corpus (Bengfort et al., 2018; Schwarz, 2019).

In this paragraph, we will look at some of the methods available in the three software platforms for calculating and visualizing word correlations.

2.6.1 Stata Code and Output

```
// Install the package from SSC (Statistical Software Component)
. ssc install txttool

// The function 'cd' changes the current working directory to the specified directory.
. cd "C:\mypath"

// Data editor example. Each row denotes a distinct piece of text
```

	text
1	The present and future of money in the digital …
2	Monetary policy and financial stability
3	Macroprudential policy for non-bank financial i…

```
// We remove punctuation and any special characters and lowercases the text. Stop words specified in a txt file are removed. The txt
file must be inside the directory specified with the 'cd' function
. txttool text, gen(bagged) stopwords("stopword.txt") stem bagwords

// Declare data to be bipartite network data (text-words)
. nw2set w*

// Display more details about the current network
. nwset, detail

// Plot the bipartite network
. nwplot, arcstyle(curved) lab

// Make one-mode projection on level 1 (words)
. nw2project networkname, level(1)
```

2.6.2 R Code and Output

```
# The line removes all variables from the current environment
> rm(list = ls())

# Loading the necessary packages
> library(dplyr)
> library(tidytext)
> library(janeaustenr)
> library(ggraph)

# Upload my dataset from a .csv format
> my_data <- read.csv("path\\tweets.csv", sep=";")

# By seeing how often word X is followed by word Y, we can build a model of the relationships between them. To tokenize by word,
we have been using the "unnest_tokens" function. To tokenize into consecutive sequences of words, we employ the "n-grams"
function. The number of words we want to capture in each n-gram is indicated by the n parameter.
> data = my_data %>%
  unnest_tokens(bigram, contents,token = "ngrams", n = 2)

# The "separate" function is used to split a column into numerous sections. This allows us to divide it into two columns, "word1" and
"word2" from which we may delete any instances where either is a stop-word. Then, we count how many times two words are
combined
> library(tidyr)
> data1=data %>%
  separate(bigram, into = c("word1", "word2"), sep = " ")
> data1 <- data1 %>%
  filter(!word1 %in% stop_words$word, !word2 %in% stop_words$word) %>%
  count(word1, word2, sort = TRUE)

# Only combinations with a frequency greater than 30 are used
> bigram_graph <- data1 %>%

  filter(n > 30) %>%
  as_tbl_graph()

# With the ggraph function, we can visualize the co-occurance network before adding layers. Three entities are added: nodes, edges,
and text
> ggraph(bigram_graph, layout = "fr") +
   geom_edge_link() +
   geom_node_point() +
   geom_node_text(aes(label = name), vjust = 1, hjust = 1)

# We add directionality with an arrow, constructed using the function grid::arrow().The arrow indicates the text's word order. We
plot the network with some options
> arrow <- grid::arrow(type = "closed", length = unit(.15, "inches"))
> ggraph(bigram_graph, layout = "fr") +
   geom_edge_link(aes(alpha = n), show.legend = T, arrow = arrow) +
   geom_node_point(color = "orange", size = 4) +
   geom_node_text(aes(label = name), size = 2,vjust = 1, hjust = 1)
```

2.6.3 *Python Code and Output*

```
# We import textnets package
>>> import textnets as tn

# Construct the corpus. From a .txt file we use tn.Corpus.from_files
>>> corpus = tn.Corpus(tn.examples)

# Textnets is deleting stop words, stemming, and removing punctuation marks, numbers, URLs, and the like because we are using
tokenized with all defaults.
>>> mytext = tn.Textnet(corpus.tokenized(), min_docs=1)

# Plot the bipartite network
>>> mytext.plot(label_nodes=True, scale_nodes_by="birank",scale_edges_by="weight")

# We can also visualize the projected network on both levels (in this case "doc" and "term")
>>> magazine = mytext.project(node_type="doc")
>>> magazine.plot(label_nodes=True)
```

2.7 Case Study Using R: Text Network on ECB Speeches for the Climate Crisis

An increasing body of economic literature applies computational linguistics methods to central bank communication. This case study contributes to research on the effectiveness of such communication (Du et al., 2018; Seelajaroen et al., 2019). Given the ECB's role—both implicit and explicit—in ensuring financial and macroeconomic stability, it has also promoted the engagement of national monetary authorities in addressing climate-related and environmental challenges at a systemic level (Carney, 2015). Today, analyzing the capacity of central banks to promote sustainable finance and address environmental risks is of strategic importance in the context of green finance governance.

Through text mining combined with network visualization, we analyze 337 ECB Executive Board speeches delivered between January 2019 and April 2022. We focus exclusively on speeches delivered in English, to demonstrate the potential of textual network analysis for extracting structured insights from policy discourse.

From a policy perspective, such analysis is particularly valuable. When climate-related financial policy is effectively communicated, these techniques can help identify and disentangle coordination problems that may arise in the implementation of monetary policy. They can also support recommendations for designing governance architectures that facilitate the transition to a low-carbon economy.

In this case study, we show how to read text data in RStudio, tokenize texts, and create a document-term matrix. The procedure we are demonstrating is useful not only for studying single words, but also for investigating word relationships. Such linkages enable us to identify which words tend to follow after others, compute co-occurences and correlations, and determine which words appear near one another. This case study also explained how to use the *ggraph* package to visualize both sorts of relationships as networks.

The first line in the next code block removes all variables from the current environment. Then, we need to get the packages we will use *tm, stringr*, *SnowballC*, *ggplot2*, *wordcloud*, *dplyr, tidytext and janeaustenr, tidyr, tidygraph*, and *ggraph*.

Input

```
> rm(list = ls())
> library("tm")
> library("stringr")
> library("SnowballC")
> library(ggplot2)
> library(wordcloud)
> library(dplyr)
> library(tidytext)
> library(janeaustenr)
> library(tidyr)
> library(tidygraph)
> library(ggraph)
```

In the next code block, we set working directory and data path. In this case, the data path is a *.csv* format composed of four columns: Date, Speaker, Title, and Text. The variables are detailed in Table 2.5.

Input

```
> setwd("C:/path/Chapter 2")
> ecb <- read.csv("C:/path/Chapter 2/Chapter 2/ECB_speeches.csv", sep=";")
```

The *grepl()* function in R searches for pattern matches within a character vector. When given a keyword, *grepl()* returns TRUE if the keyword is found in a string, otherwise it returns FALSE. This function can be used to insert a new column in the dataframe, with TRUE or FALSE indicating whether the term "climate change" is present in each row or not.

Input

```
> ecb$true<-grepl('climate change', ecb$Text)
> ecb$true
```

Table 2.5 Description of the variables

Date	The original publication date of the speech on the ECB website. Dates are given in the format YYYY-MM-DD
Speaker	List of ECB executive board speakers
Title	The title of the speech
Text	The content of each speech

The basic structure of the *tm* package is the Corpus. Corpus is a collection of text documents. *VCorpus* in *tm* refers to a "Volatile" corpus, which means that the corpus is stored in memory. To create a corpus, we need to define the source and the object. We will use the text column as the object.

We may need to invoke the function called *tm_space()* that generates content transformers once the corpus has been created. For instance, functions which modify the content of the DB object. The *gsub()* function replaces all matches of a string. The *tm_map()* function transforms the specific corpus. Any character matching the pattern specified within "" will be changed to a space. It is also possible to convert the corpus to lowercase letters, delete punctuation and English stopwords, and collapse several whitespace characters into a single blank. The final line of the box refers to the stemming method, which involves normalizing words into their base or root form.

Input

```
> DB<-VCorpus(VectorSource(ecb$Text))
> ECBSpace <- content_transformer(function(x, pattern) {return (gsub(pattern, "", x))})
> DB <- tm_map(DB, ECBSpace, "-")
> DB <- tm_map(DB, ECBSpace, "€")
> DB <- tm_map(DB, ECBSpace, ":")
> DB <- tm_map(DB, ECBSpace, "âœ")
> DB <- tm_map(DB, removePunctuation)
> DB <- tm_map(DB, ECBSpace, "’")
> DB <- tm_map(DB, ECBSpace, "‘")
> DB <- tm_map(DB, ECBSpace, "-")
> DB <- tm_map(DB, ECBSpace, "â")
> DB <- tm_map(DB, ECBSpace, "™")
> DB <- tm_map(DB, ECBSpace, "/")
> DB <- tm_map(DB, ECBSpace, "<")
> DB <- tm_map(DB, ECBSpace, "~")
> DB <- tm_map(DB, ECBSpace, "#")
> DB <- tm_map(DB, ECBSpace, "Ÿ")
> DB <- tm_map(DB, ECBSpace, "ð")
> DB <- tm_map(DB, ECBSpace, "®")
> DB <- tm_map(DB,content_transformer(tolower))
> DB <- tm_map(DB, removeNumbers)
> DB <- tm_map(DB, removeWords, stopwords("english"))
> DB <- tm_map(DB, stripWhitespace)
> DB <- tm_map(DB,stemDocument)
```

To gain a sense of the number of words in each speech, we can develop a term-document matrix. Using the *DocumentTermMatrix()* function, we are able to build and transform "DB" into a term-document matrix. With *inspect()* function, we display detailed information on our corpus.

Input

```
> dtm <- DocumentTermMatrix(DB)
> inspect(dtm[100:150, 1:50])
```

Output

```
<<DocumentTermMatrix (documents: 51, terms: 50)>>
Non-/sparse entries: 111/2439
Sparsity           : 96%
Maximal term length: 11
Weighting          : term frequency (tf)
Sample             :
     Terms
Docs abil abl absenc absorb absorpt acceler accept access accommod accompani
 101    0   3      0      0       0       0      5      1        0         0
 107    0   1      3      1       0       1      0      1        0         0
 108    1   0      0      0       0       0      2      4        0         0
 111    1   1      1      0       0       0      3      1        2         0
 126    0   0      0      0       0       1      1      2        0         4
 129    4   1      0      6       5       1      0      1        1         1
 136    0   0      0      2       0       1      2      0        1         1
 139    2   2      1      0       0       0      0      7        1         0
 144    0   1      2      1       0       1      0      0        4         0
 147    2   0      0      2       0       2      0      3        1         1
```

Out of the 337 documents in our corpus, only the first 51 are displayed for inspection. We identify the terms we want to evaluate by selecting the first 50 most frequent terms. The sparsity of the term-document matrix refers to how many terms appear in only a few documents—in this example, the sparsity is 96%, meaning most words occur in only a small subset of speeches. Additionally, the longest word identified in the corpus is composed of 11 letters.

In the following lines of code, we calculate the frequency of each word in the corpus and examine the length of the resulting "freq" matrix.

Input

```
> freq <- colSums(as.matrix(dtm))
> length(freq)
```

Output

```
[1] 8408
```

We have 8408 terms in all. The terms are then sorted in decreasing order. In the example, we request to view a sample of the most common terms. If we want to find the rarest terms, we must use *tail* rather than *head.*

Input

```
> ord <- order(freq,decreasing=TRUE)
> freq[head(ord)]
```

Output

```
bank polici financ  euro  area inflat
 3117  3102  2961  2634  2191  2082
```

There are no major surprises here. Given the context we are investigating, terms such as "bank", "policies", "finance", and "euro" are the most representative. We can also observe that certain frequent words appear more than 900 times.

Input

```
> findFreqTerms(dtm,lowfreq=900)
```

Output

```
[1] "also"     "area"     "bank"     "can"     "central" "ecb"     "econom"   "effect"   "euro"
[11] "european" "financ" "global" "increa" "inflat" "market" "measur" "monetari" "need"   "pandem"
[21] "payment"  "polici"  "price"   "rate"    "risk"    "sector"  "time"     "will"     "year"
```

The output displays a list of words whose frequencies exceed a specified threshold, but the list is not sorted. Notably, keywords related to climate change do not appear among the most frequent terms. However, we observe that the stemmed term "pandem" is among the most frequently used, which is expected given that the speeches cover the entire duration of the Covid-19 crisis. To gain a clearer understanding of how the ECB addresses climate change in its discourse, it may be useful to examine the associations between terms. This can be done using the *findAssocs()* function, which identifies terms that are statistically associated with the word "climate" across the corpus.

Input

```
> findAssocs(dtm,"climat", 0.5)
```

Output

```
$climat
   relate  chang       greenhous    disclosur   neutral  strand
    0.77   0.73          0.69          0.63       0.59    0.58
   emiss  prejud sustainability carboninten   transit  carbonneutr pollut
    0.54   0.54          0.54          0.53       0.52        0.51    0.51
```

In this case, we have a list of words that are connected with "climat" with a correlation greater than 0.50. The greater the coefficient we specify, the fewer the linked terms. With the highest correlations we find relate (0.77), change (0.73), and greenhouse (0.69).

In the following code, we generate a dataframe with the words and their frequencies. Then, we visualize a barplot (Fig. 2.12) using the *ggplot2* library. We have decided to display only terms with a frequency greater than 900.

Input

```
> wf=data.frame(term=names(freq),occurrences=freq)
> p <- ggplot(subset(wf, freq>900), aes(term, occurrences))
> p <- p + geom_bar(stat="identity", fill="steelblue") + theme_classic()
> p + theme(axis.text.x=element_text(angle=45, hjust=1))
```

Output

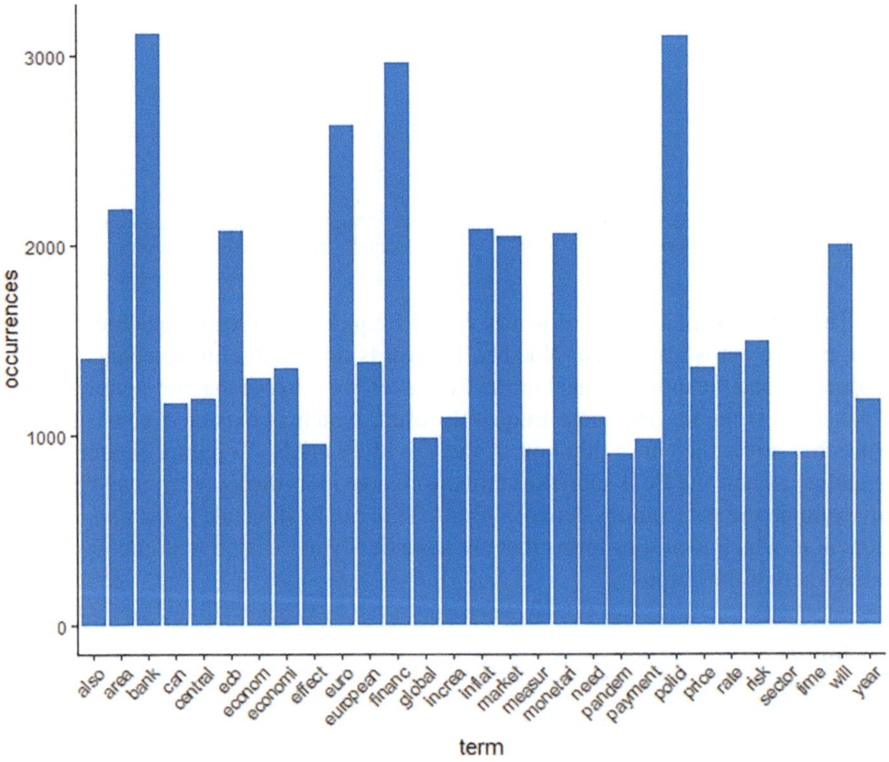

Fig. 2.12 Barplot of main words and their frequencies

Finally, we create a word cloud, also referred to as a text cloud, which is a specific type of visual representation for text data.

Input

```
> wordcloud(names(freq),freq,min.freq=900,colors=brewer.pal(6,"Dark2"))
```

Output

It is important to note that the modal verb "will" predominates in the corpus. This could signal that the ECB intends to take measures or make decisions on certain issues in the near future, indicating that the speeches are mostly concerned with the future. In the following code block, we create a function that searches for the keyword "climat" in each text string and then assigns a category in an R dataframe (output in Fig. 2.13).

Input

```
> getCount <- function(data,keyword)
{
> wcount <- str_count(ecb$Text, keyword)
> return(data.frame(data,wcount))
}
> climat<-getCount(ecb$Text,'climat')
```

Output

	data	wcount
1	Fostering resilience and convergence in Economic and Mo...	0
2	Tenth anniversary of the euro in Slovakia Speech by Yves ...	0
3	Debate about the Annual Report Speech by Mario Draghi...	0
4	20th anniversary of the euro Speech by Mario Draghi, Pre...	0
5	Speech A supervisory perspective on 2019 and beyond ...	1
6	Hearing of the Committee on Economic and Monetary Aff...	0
7	The possible triangle: frictionless movement of payments, ...	0
8	Fintech for the people Keynote speech by BenoÃ®t CÅ"u...	0
9	The changing role of central banking Speech by Yves Mer...	0
10	Promoting innovation and integration in retail payments t...	0
11	The euro area: current status and the monetary policy stan...	0

Fig. 2.13 "Climat" count across rows

From the output, we can see that the word "climat" appears in only one of the first 11 speeches. When we examine the full corpus, there are only a few occurrences of the keyword "climat", as is the case for many terms associated with the climate crisis.

Until now, we have examined the frequency of words in the corpus, as well as the frequency and presence of a word like "climate" in each speech. Now, we will launch the process for analyzing word pairs that will form our textual network. We aim to identify which words tend to follow others immediately, or co-occur within the same speech. To start, we split a column into tokens, flattening the table into one token per row. Tokenization essentially refers to breaking up a larger body of text into single units, in this case, words.

Input

```
> ecb = ecb %>%
  unnest_tokens(bigram, Text,token = "ngrams", n = 2)
```

Recall that the operator %>% is known as the forward pipe operator. It is used to express a sequence of operations. To tokenize by word, we use the function *unnest_tokens()*, and *token = "ngrams"* to indicate that we want to tokenize into consecutive sequences of words, in this case two. A column called *bigram* containing all consecutive word pairs will be added to our data frame *ecb*. Next, we create a new data frame called *ecb1* in which we separate the *bigram* column into two new columns, *word1* and *word2*. This allows us to remove cases where either column contains a stop-word. Lastly, we apply stemming, and remove punctuation, numbers, single words, and words with three or more repeating characters.

Input

```
> ecb1=ecb %>%
  separate(bigram, into = c("word1", "word2"), sep = " ")
> ecb1 <- ecb1 %>%
  filter(
> !str_detect(word1, pattern = "[[:digit:]]"),  # removes any words with numeric digits
> !str_detect(word1, pattern = "[[:punct:]]"), # removes any remaining punctuations
> !str_detect(word1, pattern = "(.)\\1{2,}"),  # removes any words with 3 or more repeated letters
> !str_detect(word1, pattern = "\\b(.)\\b"),   # removes any remaining single letter words
> !str_detect(word2, pattern = "[[:digit:]]"), # removes any words with numeric digits
> !str_detect(word2, pattern = "[[:punct:]]"),  # removes any remaining punctuations
> !str_detect(word2, pattern = "(.)\\1{2,}"),  # removes any words with 3 or more repeated letters
> !str_detect(word2, pattern = "\\b(.)\\b")     # removes any remaining single letter words
  )
# add stemming process
> ecb1 <- ecb1 %>%
  mutate(word = corpus::text_tokens(word1, stemmer = "en") %>% unlist())
  mutate(word = corpus::text_tokens(word2, stemmer = "en") %>% unlist()) %>%
# remove stop words
filter(!word1 %in% stop_words$word,
       !word2 %in% stop_words$word) %>%
count(word1, word2, sort = TRUE)
```

In the next code block, we define the function that allows us to visualize the text network of our corpus (Fig. 2.14). The graph in this example is undirected, since we do not account for the sequence of word pairs. We also apply a filter to display only the most frequent co-occurrences—those appearing more than 150 times.

Input

```
> graph <- ecb1 %>%
  filter(n > 150) %>%
  as_tbl_graph()

> graph

> ggraph(graph, layout = "fr") +
  geom_edge_link() +
  geom_node_point() +
  geom_node_text(aes(label = name), vjust = 1, hjust = 1)
```

Output

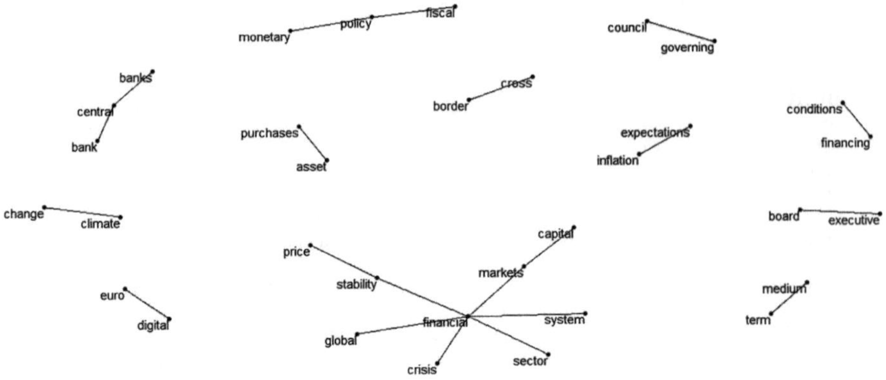

Fig. 2.14 Text network visualization

From the graph above, it can be seen that the term "financial" is the most central. It is linked with nouns like "stability", "global", "crisis", "sector", "system", and "market". There are no combinations that link to the climate crisis when the most common pairings are examined. Now, we want to see the word sequence, specifically which word comes first. We will also add color to our nodes (words) and directed links (Fig. 2.15).

Input

```
> arrow <- grid::arrow(type = "closed", length = unit(.15, "inches"))
> ggraph(bigram_graph, layout = "fr") +
  geom_edge_link(aes(alpha = n), show.legend = F,
  arrow = arrow, end_cap = circle(0.07, "inches")) +
  geom_node_point(color = "steelblue", size = 5) +
  geom_node_text(aes(label = name), vjust = 1, hjust = 1)
```

Output

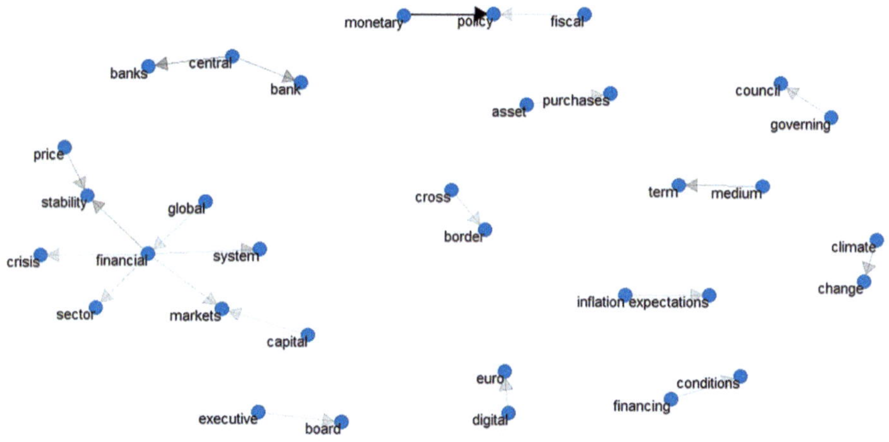

Fig. 2.15 Word sequence direction

In this case, we are working with a directed graph, where arrows indicate the direction of word sequences. A word is considered to precede another if there is an outgoing edge pointing to it. For example, "financial" precedes words such as "system" and "sector", but not "global". The intensity of the arrow color reflects the weight of the connection, with darker shades indicating stronger associations.

We then broaden our analysis by lowering the frequency threshold to include word pairings that occur more than 50 times, in order to assess whether any terms related to climate change emerge in the network (Fig. 2.16).

Input

```
> ecb1 %>%
  filter(n > 50) %>%
  as_tbl_graph() %>%
  ggraph(layout = "fr") +
  geom_edge_link(aes(alpha = n), show.legend = FALSE,
           arrow = arrow, end_cap = circle(0.07, "inches")) +
  geom_node_point(color = "orange", size = 5) +
  geom_node_text(aes(label = name), size = 4, vjust = 1, hjust = 1,check_overlap = TRUE)
```

Output

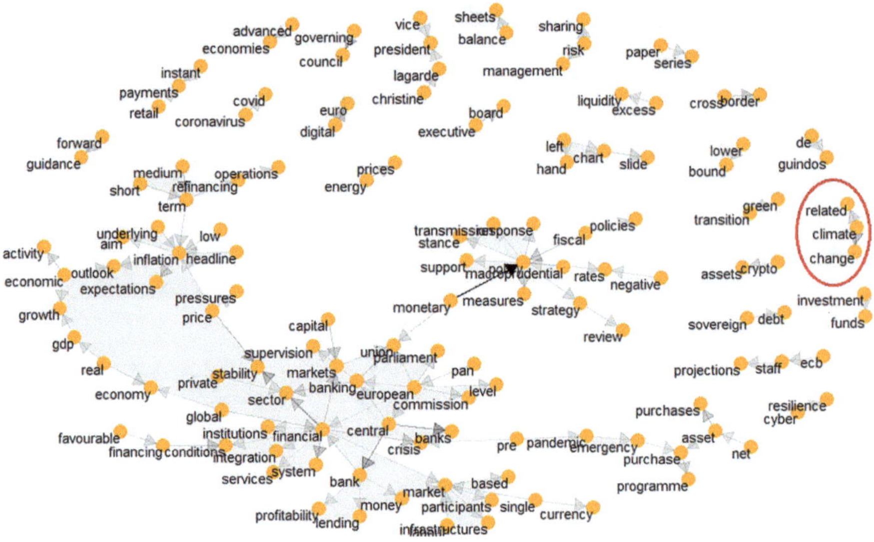

Fig. 2.16 Word sequence direction

Finally, if we want a list of words associated with the word 'climate', we need to run the following code. In this case, 'climate' appears in the first column; to find connections where 'climate' is in the second column, we need to switch from word1 to word2. (output in Table 2.6).

Input

```
> ecb2 %>%
  separate(bigram, into = c("word1", "word2"), sep = " ") %>%
  filter(word1 == "climate") %>%
  count(word1, word2, sort = TRUE)
```

Output

Table 2.6 List of words connected with the word "climate"

word1	word2	n
climate	act	1
climate	action	1
climate	agency	1
climate	agenda	1
climate	agreement	1
climate	ambition	1
climate	catastrophe	1
climate	challenge	1

In this case, we only see the first eight word combinations containing climate in the first position. The frequency of the pair in the corpus is indicated in the third column.

2.8 Conclusions

We have examined whether European Central Bank speeches addressed the climate crisis. Based on a sample of speeches held by Executive Board members between January 2019 and April 2022, we noticed that the number of keywords associated with the climate crisis is really low. We also know that communicating about the climate crisis becomes more difficult during the Covid-19 pandemic. This is evident in some of the terms associated with the epidemic itself. It will be interesting to observe if the climate crisis becomes a focal point in ECB speeches after the pandemic period.

References

Al-Taie, M. Z., & Kadry, S. (2017). *Python for graph and network analysis* (pp. 1–184). Springer.

Bengfort, B., Bilbro, R., & Ojeda, T. (2018). *Applied text analysis with python: Enabling language-aware data products with machine learning*. "O'Reilly Media, Inc".

Borgatti, S. P., Everett, M. G., Johnson, J. C., & Agneessens, F. (2022). *Analyzing social networks using R*. Sage.

Carney, M. (2015). *Breaking the tragedy of the horizon—Climate change and financial 18 stability (speech)*. Bank of England, Speech.

Du, B., Fung, S., & Loveland, R. (2018). The informational role of options markets: Evidence from FOMC announcements. *Journal of Banking & Finance, 92*, 237–256.

European Central Bank. (25 October 2019). *Speeches dataset*. Retrieved from: https://www.ecb.europa.eu/press/key/html/downloads.en.html

Kolaczyk, E. D., & Csárdi, G. (2014). *Statistical analysis of network data with R* (Vol. 65). Springer.

McKinney, W. (2012). *Python for data analysis: Data wrangling with Pandas, NumPy, and IPython*. "O'Reilly Media, Inc".

Opsahl, T. (2013). Triadic closure in two-mode networks: Redefining the global and local clustering coefficients. *Social Networks, 35*(2), 159–167.

Schwarz, C. (2019). Lsemantica: A command for text similarity based on latent semantic analysis. *The Stata Journal, 19*(1), 129–142.

Seelajaroen, R., Budsaratragoon, P., & Jitmaneeroj, B. (2019). Do monetary policy transparency and central bank communication reduce interest rate disagreement? *Journal of Forecasting, 39*(3), 368–393.

Zinilli, A., & Cerulli, G. (2017). Datanet: A Stata routine for organising a dataset for network analysis purposes. *International Journal of Computational Economics and Econometrics, 7*(4), 454–462.

Chapter 3
Network Metrics

3.1 Introduction

Following the formalization of the algebraic structure that allows us to represent the network in mathematical terms, it is feasible to develop various measurements that can highlight particular characteristics of the network under investigation. In this section, some of the most typical measures and techniques used in social network analysis are illustrated. The many structural analysis concepts and techniques addressed in this chapter have been arranged, for ease of presentation, into two main categories: (i) interpretation of vertices and edges and (ii) interpretation network cohesiveness (i.e., involving a larger set of vertices and edges). Please note that many additional concepts and methods require a depth of understanding beyond the scope of this chapter. Furthermore, keep in mind that different software may select and implement centrality measures in different ways, making them not exactly identical.

3.2 Paths

A component is a part of the graph comprising all vertices that form a connected subgraph (Fig. 3.1). It is possible to have:

- A major (also called main or giant) component, which is the largest component within a network.
- A minor component, which is smaller than the main component. A network typically contains multiple minor components.
- An isolated vertex, namely a vertex that has no edges to the other vertices within the graph.

A. Zinilli, *Elements of Network Science*, Statistics and Computing,
https://doi.org/10.1007/978-3-031-84712-7_3

Figure 3.1 shows the various components and isolated vertices. Each pair of vertices within the components is connected by a path, which is a set of vertices with the peculiarity that every succeeding pair of vertices in the set is linked by an edge. When two vertices in a network are connected, the path with the fewest edges is known as the shortest path or geodesic path. The term "geodesic distance" or "shortest distance" refers to the length of a geodesic path. Although geodesic pathways are not always distinct, the geodesic distance is clearly defined since they are all the same length. The largest geodesic distance in the graph, the diameter of the network may also be considered. However, since it only measures the distance between one pair of vertices at the extreme end of the range of distances, the diameter is a less meaningful measurement for real networks than the average distance. Additionally, the diameter is sensitive to relatively slight changes in the list of vertices, making it a subpar predictor of how the network will behave overall.

The graph in the next illustration (Fig. 3.2) has 5 vertices and 6 edges in total.

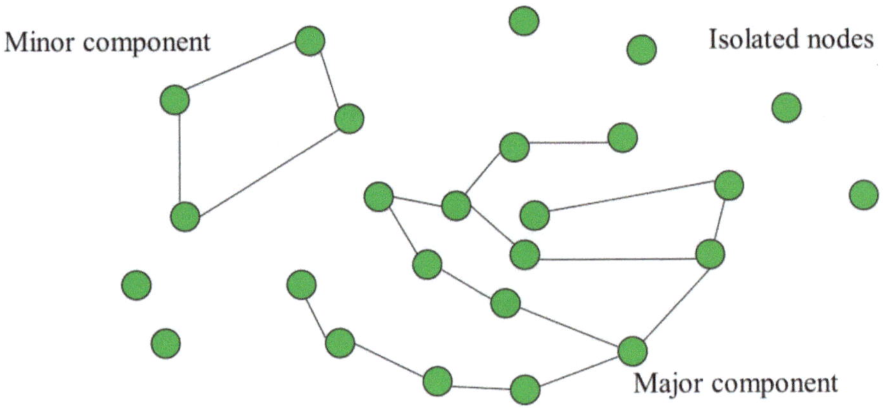

Fig. 3.1 Network components

Fig. 3.2 Network components

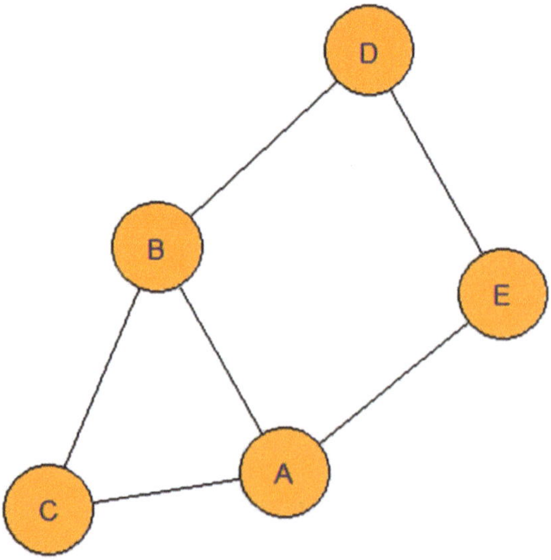

Focusing on a randomly selected pair of vertices such as *C* and *D* in the Fig. 3.2, we notice that between this pair of vertices there are different paths (*C-B-D; C-A-B-D; C-A-E-D*). The shortest path between *C* and *D* is the geodesic distance, which is in this example equal to *C-B-D* (two edges from *C* to *D*).

A cycle is a significant form of path; it represents a ring-like structure, exemplified by vertex sequences like *A-B-C-A* or *A-B-D-E-A* in Fig. 3.2. A cycle, as shown in the example, is a path with at least three edges in which the first and last vertices of the path sequence are the same but all other vertices in the sequence are distinct.

3.3 Density of Graphs

Networks can be studied in a variety of ways using different and specific methods. One method is to examine their structural characteristics, which include size, density, topology, and statistical properties. Size and density are the most basic network measures, which are highly relevant for comparing multiple networks. It is worth noting that a network's size and density do not reveal much about its actual topology. Many networks with different topologies can have the same size and density. The size of a network is characterized by its number of vertices; while the level of connectivity in the network is described by its number of edges (links). As mentioned in Chap. 2, the total number of possible edges in an undirected complete network with *N* vertices, where loops (edges that connect a vertex to itself) are not allowed and the direction of a relationship is irrelevant, is calculated as:

$$E_{tot} = \frac{N(N-1)}{2} \tag{3.1}$$

For example, consider a network with five vertices (e.g., people or organizations) where each vertex is assumed to be connected to every other vertex. In this case, the network would have:

$$E_{tot} = \frac{5(5-1)}{2} = 10 \text{ total edges} \tag{3.2}$$

In a directed network with *N* vertices, however, we would have:

$$E_{tot} = N(N-1) \tag{3.3}$$

In the same example, but with the network now directed, we would have:

$$E_{tot} = 5(5-1) = 20 \text{ total edges} \tag{3.4}$$

In real-world networks, *E* is often much smaller than E_{tot}, indicating that most real networks are sparse. Density is a network property that measures group cohesiveness. It is simply the ratio of observed edges to the total number of possible edges. This metric assumes a minimum value of 0 when the network consists of isolated vertices (i.e., no edges), and a maximum value of 1 when each vertex is connected to all the others.

3.3.1 Stata Code and Output

```
// We use the Zachary's Karate Club graph
. clear all
. nwset, nwclear

// Uploading a network from other sources: UCINET
. nwimport http://vlado.fmf.uni-lj.si/pub/networks/data/ucinet/zachary.dat, type (ucinet)

// Description of the network
. nwsummarize, detail

// Getting a table that lists all the dyads in the network and displays the distribution of dyad values
. nwtabulate ZACHC

// Getting the number of mutual, asymmetric, and null dyads, as well as the reciprocity coefficient
. nwdyads ZACHC

// Assembling a table with information about each type of triad in the network
. nwtriads ZACHC

// Simmelian ties
. nwsimmelian ZACHC

// Plotting the Simmelian ties
. nwplot ZACHC, edgecolor (_simmelian)
```

Output

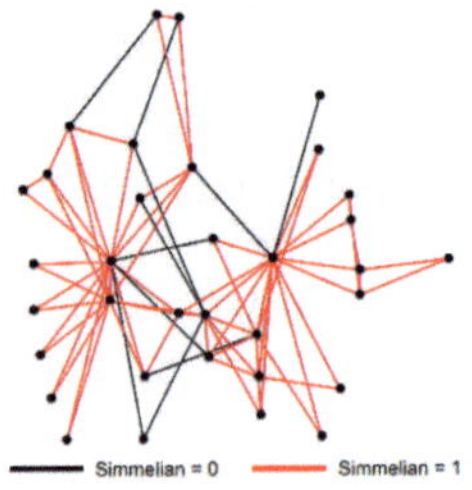


```
// Getting network components (subgraph of connected nodes)
. nwcomponents ZACHC

// The following command gives information on total paths, diameter and average shortest path
. nwgeodesic ZACHC

// Getting the number of bridges in the network (a bridge is an edge that does not belong to a triangle)
. nwbridge ZACHC, detail

// Getting the local bridges in the network (a local bridge is an edge that reduces the distance between a pair of nodes)
. nwbridge ZACHC, detail local
```

Output

```
Directed     : false
Bridges      : 1
Local bridges: 10
-----------------------------

   Detailed bridges (local):
   Span in parenthesis

    (var1 <=> var32) (3)
    (var2 <=> var31) (3)
    (var3 <=> var10) (3)
    (var3 <=> var28) (3)
    (var3 <=> var29) (3)
    (var10 <=> var34) (3)
    (var14 <=> var34) (3)
    (var20 <=> var34) (3)
    (var24 <=> var26) (3)
    (var25 <=> var28) (3)
```

The **span** of a local bridge is the distance between its ends if we remove the arch. The greater the span the more important the cohesion role of the arch.

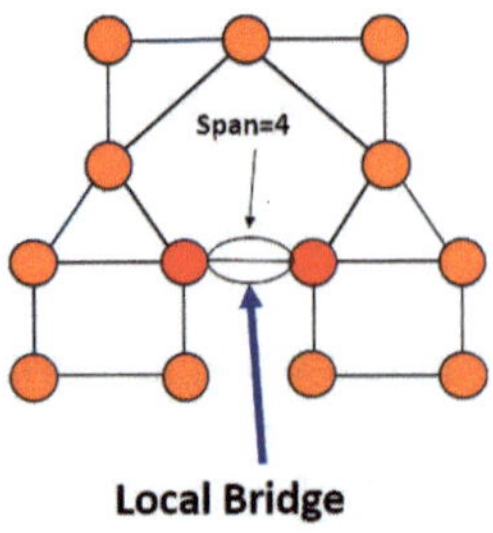

3.3.2 R Code and Output

```
# Downloads the Zachary's Karate Club graph
> install.packages("CINNA")
> library(CINNA)
> data("zachary")
> print(zachary)

# Loading package igraph and network
> library(igraph)
> library(network)

# Now let us duplicate the Zachary dataset and give it the name 'g'
> g <- zachary

# Count number of edges
> gsize(g)
> gorder(g)


# Identifying the two nodes in the graph that are most distant from one another
> farthest_vertices(g)
> get_diameter(g)

# Density calculation
> density<- graph.density(g, loop=FALSE)

# Get the diameter of the graph 'g'
> diam<-diameter(g, directed = FALSE)

# Get the average path length of the graph 'g'
> averagepath <- mean_distance(g, directed = FALSE)
> averagepath

# Find triangles in the graph
> count_triangles(g)

# Display the first 20 triangles in the network
> triangles <- matrix(triangles(g))

# Find out which cliques are the biggest in the network
> largest_cliques(g)

# Identify all network maximal cliques and assign them to the object "cliques"
> cliques <- max_cliques(g)

# Determine the dimensions of each maximal clique
> table(unlist(lapply(cliques, length)))
```

3.3.3 *Python Code and Output*

```
# Downloads the necessary libraries
>>> import matplotlib.pyplot as plt
>>> import networkx as nx
>>> import pandas as pd
>>> import numpy as np

# We use the Zachary's Karate Club graph
>>> g = nx.karate_club_graph()
>>> nx.draw(g)

# List the isolated nodes in the graph.
>>> list(nx.isolates(g))

# Determines whether the node 3 is an isolate [True or False].
>>> nx.is_isolate(g, 3)

# Calculating the shortest paths of all combinations of nodes
>>> p = nx.shortest_path(g)
>>> print(p)

# Returns average of shortest paths between all possible pairs of nodes
>>> print(nx.average_shortest_path_length(g))

# Returns the Eccentricity, Diameter and Radius of the graph. The eccentricity is the largest possible shortest path distance between
node i and all other nodes. The Diameter is the maximum shortest distance between a pair of nodes in a graph. It is the largest
possible eccentricity value of a node. The Radius is the minimum eccentricity value of a node.
>>> print("Eccentricity: ", nx.eccentricity(g))
>>> print("Diameter: ", nx.diameter(g))
>>> print("Radius: ", nx.radius(g))

# Generate all bridges in a graph.
>>> list(nx.bridges(g))

# Compute the number of triangles.
>>> print(nx.triangles(g))

# Connected components are sorted in descending order of their size
>>> g_components = nx.connected_component_subgraphs(g)
>>> g_mc = g_components[0]
```

3.4 Centrality Measures

Identifying the most central nodes in a network is one of the key goals of Network Science, especially at the exploratory stage. The term "central" refers to a set of metrics that, when taken together, indicate the importance of nodes within a network. These metrics are initially specified at the level of individual nodes but can be expanded to include recognizable groupings of nodes within the network under study. The centrality of a node in Network Science is a measure of its relative importance within a graph showing the quantity and quality of a node's connections with others. Centrality measures can help in identifying popularity, most liked, and biggest influencers within the network. Centrality metrics have a wide range of applications, including: (i) identifying important players in social networks, such as opinion leaders or influential people; (ii) determining the resilience of infrastructure networks, such as electricity grids or transportation systems to deliberate attacks or malfunctions; (iii) ranking websites or academic works according to their popularity or impact; (iv) examining the structure of financial networks, such as banking systems or stock markets, and identifying which nodes are the most connected or risky; (v) assessing the effectiveness of communication networks, such as the internet or phone networks, and identifying which points are crucial for the effective transmission of information. In general, centrality measures allow researchers and practitioners to understand the composition and operation of complex systems as well as to create efficient plans for improvement, regulation, or intervention. In order to prevent oversimplifications or misunderstandings, it is crucial to remember that centrality measures should be used in conjunction with other network metrics and contextual information. Here we will discuss some of the most commonly used centrality measures.

3.4.1 Degree Centrality

The primary measure of centrality is degree centrality, which states that a node is central in the network if it is among the most active nodes i.e., if it has significantly more connections than others. A node's degree is simply the number of arcs that connect it to other nodes. The number of edges that occur to a node i within the network under investigation determines its degree, or k_i. A node's degree can be expressed equivalently as the cardinality of its neighborhood, or the number of nodes that are immediately next to it. In the case of direct networks (edges with associated directions), the degree differs into in-degree and out-degree. The in-degree is the number of bonds directed to the node, while the out-degree is the number of bonds directed to others.

For a given graph $G = (V, E)$ with $|V|$ vertices and $|E|$ edges, the degree centrality of a vertex v is defined as:

$$C_D(v) = \deg(v) \rightarrow C_D(i) = \sum_{j=1}^{N} a_{ij} \rightarrow k_{(i)} = \sum_{j=1}^{N} a_{ij} \tag{3.5}$$

Let a_{ij} be the observed value in the adjacency matrix A and it is equal to 1 if there is an edge from node i toward the node j, and $a_{ij} = 0$ otherwise. Here, N denotes the set of nodes, thus $i, j \in \{1,\ldots,N\}$.

For instance, in Fig. 3.3 below, the degree of node A (in absolute value) is equal to 3.

Typically, the degree centrality of nodes is normalized by considering its highest value in the observed graph, which for the degree centrality is equal to N-1. In Fig. 3.1, the normalized degree of node A is equal to 0.75, and it is given by:

$$C_D^{norm}(i) = \frac{a_i}{N-1} \rightarrow k_i^{norm} = \frac{a_i}{N-1} \tag{3.6}$$

Fig. 3.3 Graph example

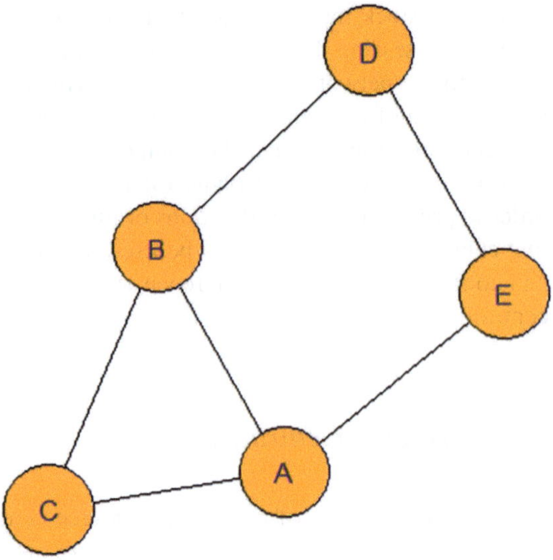

For directed networks, it is essential to distinguish between incoming and outgoing edges associated with a node. We distinguish between in-degree and out-degree:

$$C_D(i)^{IN} = \sum_{j=1}^{N} a_{ji} \rightarrow k_i^{in} = \sum_{j=1}^{N} a_{ji} \tag{3.7}$$

$$C_D(i)^{OUT} = \sum_{j=1}^{N} a_{ij} \rightarrow k_i^{out} = \sum_{j=1}^{N} a_{ij} \tag{3.8}$$

Network centralization is the term used when the notion of centrality at the node level is applied to the entire graph. Let v^* stand for the vertex in G that has the greatest degree of centrality. Let U^* be the node in X with the highest degree centrality, and let $X = (Y,Z)$ be the $|Y|$ node linked network that maximizes the following quantity:

$$H = \sum_{j=1}^{|Y|} \left[D(y^*) - D(y_i) \right] \tag{3.9}$$

When the network X has a single central vertex to which all other vertices are linked, the value of H reaches its maximum. It follows that the graph G is centralized to the following degree:

$$D(G) = \frac{\sum_{i=1}^{|V|} \left[D(v^*) - D(v_i) \right]}{H} \tag{3.10}$$

The basic topological description of a network may be expressed in terms of the degree distribution *P(k)*, which is defined as the probability that a node randomly selected has degree *k*, or, alternatively, as the proportion of nodes in the network that have degree *k*. In the case of direct networks, the two distributions $P(k_{in})$ and $P(k_{out})$ must be considered. Plotting the distribution *P(k)*, $P(k_{in})$, and $P(k_{out})$ or computing the distribution's moments might provide information on how the degree is spread across nodes in a network. The n_{th} moment of *P(k)* is defined as:

$$\langle k^n \rangle = \sum_k k^n P(k) \tag{3.11}$$

The first moment is the average degree of P(k) $G\langle k \rangle = \sum_k kP(k) = \frac{1}{N}\sum_i k_i = \frac{2L}{N}$. The second moment measures the degree distribution's variability. In Chap. 4, we will address specific aspects of the degree distribution properties.

3.4.2 Node Strength

For an undirected network, the *W* matrix defines the strength s_i of a node as the sum of the weights of the edges incident on a node as follows:

$$s_i = \sum_{j \in \mathbb{N}} w_{ij} \tag{3.12}$$

If the network is directed, there are two factors that determine the node strength: the quantity of incoming (s_i^{in}) and outgoing (s_i^{out}) weighted edges:

$$\begin{aligned} s_i^{in} &= \sum_j w_{ji} \\ s_i^{out} &= \sum_j w_{ij} \\ s_i &= \sum_{j \in \mathbb{N}} w_{ij} \end{aligned} \tag{3.13}$$

Total strength is finally defined as:

$$s_i = s_i^{in} + s_i^{out} \tag{3.14}$$

Similar to the degree, we may obtain the strength distribution *P(s)*. In the case of directed networks, it is essential to consider two distributions $P(s_i^{in})$ and $P(s_i^{out})$. Nodes with significantly greater strength than the rest of the network's nodes are referred to as hubs.

3.4.3 Closeness Centrality

The degree centrality measure only captures a node's direct interactions with other nodes nearby, not relationships at subsequent depth levels. In general, the dispersion of information in connected graphs is modeled using minimum pathways. The closeness centrality measure highlights the distance between one node and all others in the graph. Closeness centrality is defined as the average distance between a starting node and all other nodes in the network. In an undirected graph, the Closeness Centrality for generic node *i* is given by:

$$C_c(i) = \left[\sum_{j=1}^{N} d(i,j) \right]^{-1} \tag{3.15}$$

It is the inverse of the sum of the geodesic distances between a node *i* and all the other nodes in the graph; this index has a maximum value of $(N - 1)^{-1}$ when the node is next to all the others. The inverse is used so that higher values indicate more central nodes.

Figure 3.4 displays an example graph with 5 nodes, illustrating the Closeness calculation specifically for node *A* (Closeness Centrality equal to 0.8).

Fig. 3.4 Graph example

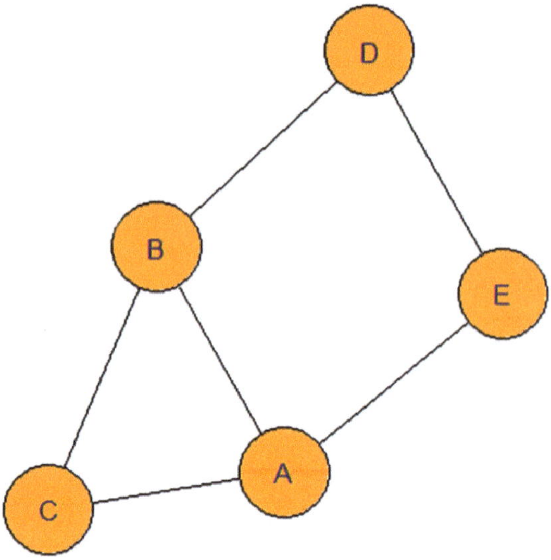

$$C_c(A) = \left[\frac{\left[\sum_{j=1}^{N} d(A,j)\right]}{N-1}\right] = \left[\frac{d(AB)+d(AC)+d(AD)+d(AE)}{N-1}\right]^{-1}$$
$$= \left[\frac{1+1+2+1}{5-1}\right]^{-1} = 0.\overline{8} \tag{3.16}$$

This kind of measure is very helpful for figuring out which nodes can transmit information more quickly than others. Closeness centrality in directed networks varies based on the direction of the edges connecting two nodes:

- Out-closeness centrality measures how close a node is to the nodes it sends information to.
- In-closeness centrality measures how close a node is to the nodes it receives information from.

3.4.4 Betweenness Centrality

Betweenness centrality is a metric that may be calculated both at the node and edge levels, and it reflects the likelihood that a node (or edge) lies along the shortest path between two other nodes. More specifically, the node with a high betweenness is known as the broker, and the bridges are the edges that form the connections.

Betweenness centrality holds that a node is considered central if it frequently acts as an intermediary between other nodes that would otherwise be unconnected.

Mathematically, we define betweenness in this way:

$$C_B(i) = \sum_{j<k} g_{jk}(i) / g_{jk} \tag{3.17}$$

where $g_{jk}(i)$ represents the number of shortest paths between nodes j and k with node i acting as an intermediate node in the shortest path.

In this simple example (Fig. 3.5), there are no alternate paths.

A lies between no two other vertices.
B lies between *A* and 3 other vertices: *C*, *D*, and *E* - (*AC*), (*AD*), (*AE*).
C lies between 4 pairs of vertices: *A*, *B*, *D*, *E* - (*AD*), (*AE*), (*BD*), (*BE*).
D lies between *E* and 3 other vertices: *A*, *B*, and *C* - (*AE*), (*BE*), (*CE*).
E lies between no two other vertices.

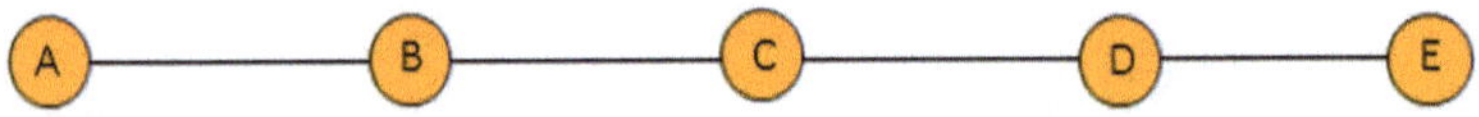

Fig. 3.5 Graph example

We can conclude that C gets full credit.

The capacity of intermediation places the node in an advantageous position because, by definition, the flow of information between two agents always takes the shortest path, this node represents a mandatory step between two nodes. When calculating the betweenness centrality, our main goal is to identify the nodes that connect the different components and that are crucial for information flow as well as the overall connectivity of different components of the same network.

3.5 Spectral Centrality Measures

Due to the heterogeneous distribution of a node's neighbors, nodes with low degree and nodes with high degree can coexist in many networks. Real networks have a number of outstanding features that can be attributed to this heterogeneity, including resilience to random failures and attacks, as well as the propagation of innovation and knowledge (Krackhardt & Stern, 1988; Barabási, 2016; De Domenico & Arenas, 2017). Some metrics of centrality are dependent on the spectral structure of the graph matrices, e.g., information we can derive from the eigenvalues and eigenvectors of the matrices. These measures are typically used on directed graphs, but when appropriate, can be extended to the undirected cases. In this section, we will introduce and discuss four centrality measures: Jaccard coefficient, Eigenvector centrality, Katz centrality, and PageRank centrality. Before introducing these measures and providing some insights for them, let us begin by specifying the concept of neighborhood.

3.5.1 Neighborhood

A node's neighborhood includes the group of nodes to which it is connected. Studying the neighborhood of a given node i aids in identifying vertices, or communities of vertices, which share specific characteristics. When individuals are motivated to emulate the behavior of their neighbors, cascading effects can occur. These effects can be observed when a new behavior initiates with a small group of early adopters and then spreads radially outward through the network.

$$N_i(G) = \{j : g_{ij} = 1\} \tag{3.18}$$

The neighborhood of i in graph G exists only if ij is equal to 1 (indicating the presence of an edge).

The neighborhood degree sequence for the node i, s_i, is derived from:

$$s_i = \left\{k_1^i, k_2^i k_3^i \ldots, k_n^i\right\} \tag{3.19}$$

with k_i equal to the degree of the nodes to which i is connected. Being a sequence of degrees:

$$k_1^i \leq k_2^i \leq k_3^i \ldots \leq k_n^i \tag{3.20}$$

Analysis of the degree sequences of neighbors of a generic node i may help in examining the variance of neighborhood degree sequences in the context of hierarchical complexity. Additionally, the neighborhood degree sequence may analyze neighbors of the same degree to determine how similar the neighbors of node i are.

3.5.2 *Jaccard Coefficient*

The Jaccard coefficient is used to calculate how similar the neighbors of two vertices are. The vertices that are directly connected to a specific vertex are referred to as that vertex's neighbors. The Jaccard coefficient, which is frequently employed in information retrieval, is used to determine the proximity between two vertices. It may be calculated using the following equation:

$$JC = \frac{\left|N_i \cap N_j\right|}{\left|N_i \cup N_j\right|} \tag{3.21}$$

where N_i is the list of neighbors of i, and N_j is the list of neighbors of j. As a result, the numerator represents the number of common neighbors, whereas the divisor is the number of distinct i and j neighbors.

The Jaccard coefficient provides a measure that indicates how similar two vertices' neighbors are. Given two vertices N_i and N_j, the Jaccard coefficient is the ratio of the size of the intersection to the size of the union of their respective neighbors. Consider the following simple example:

Antonio = [Claudia, Anna, Jimmy, Robert, Will].
Valentina = [Claudia, Robert, Will, Jennifer, Elizabeth].

Taking into account the neighbors in square brackets, we wish to determine how similar the neighborhoods of the nodes Antonio and Valentina are. To calculate the Jaccard coefficient of the two nodes Antonio and Valentina, we first count the total number of shared neighbors, then divide by the total number of unique members.

Number of shared members (intersection): [Claudia, Will, Robert] = 3.
Total distinct members (union): [Claudia, Anna, Jimmy, Robert, Will, Jennifer, Elizabeth] = 7.
Jaccard coefficient (similarity) = 3/7 = 0.43.

The Jaccard coefficient is a value between 0 and 1, where 0 denotes the maximum dissimilarity and 1 denotes the largest similarity between two nodes.

Finally, the Jaccard distance is the complement of the Jaccard coefficient; in our simple example, the Jaccard distance is (1–0.43) = 0.57.

3.5.3 Eigenvector Centrality

Eigenvector centrality is a measure of node importance in a network that is a logical progression from simple degree centrality. Unlike degree centrality, which counts the number of a node's direct connections, eigenvector centrality considers both the quantity and quality of a node's connections. This means that a node is considered more important if it is connected to other important nodes. The centrality score is determined by the node's position in the eigenvector associated with the largest eigenvalue of the adjacency matrix of the network. In this sense, the concept assumes that important nodes serve as hubs in the network. While the number of connections a node has can contribute to its importance, the significance of those connections can matter even more: a node connected to several high-importance nodes might be deemed more important than one with many connections to less important nodes, depending on the overall network structure. Mathematically, a node's eigenvector centrality is determined by the eigenvector associated with the largest eigenvalue (λ) of the network's adjacency matrix. Specifically, if A is the adjacency matrix and x is the centrality vector, the equation $Ax = \lambda x$ is used to compute the eigenvector centrality. Thus, each component of x (i.e., the centrality of each node) is proportional to the sum of the centralities of its neighbors. Nodes with high eigenvector centrality are often connected to other nodes with high eigenvector centrality. This statistic can be used to identify clusters of highly connected and influential nodes within the network. The eigenvector is given by

$$C_E(i) = \frac{1}{\lambda_{\max}} \sum_{j=1}^{N} a_{ij} x_j \tag{3.22}$$

where λ is the maximum eigenvalue for an adjacency matrix A, a_{ij} indicates the elements of the adjacency matrix, and x_j is the eigenvector associated with the eigenvalue. Eigenvector centrality can be calculated for both undirected and directed networks. Either directed or undirected networks can theoretically have their eigenvector centrality determined. However, the interpretation and computation are often more straightforward in the case of undirected networks due to the symmetry of the adjacency matrix. In most cases, particularly for connected networks, all vertices'

$$A = \begin{pmatrix} 0 & 1 & 1 \\ 1 & 0 & 1 \\ 1 & 1 & 0 \end{pmatrix}$$

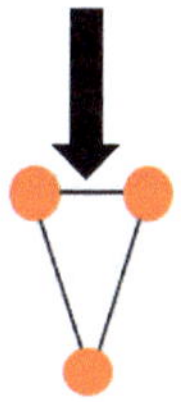

Fig. 3.6 Adjacency matrix A

eigenvector centralities are positive. For a more detailed discussion, see appropriate references such as works by Newman (2018). Here, we illustrate how to calculate the eigenvector of a simple adjacency matrix and its related eigenvalues. Suppose we have the adjacency matrix A (Fig. 3.6) composed of three nodes:

The eigenvectors and eigenvalues are calculated from the λ matrix A, which is obtained by subtracting λ from the diagonal.

$$A_\lambda = \begin{pmatrix} -\lambda & 1 & 1 \\ 1 & -\lambda & 1 \\ 1 & 1 & -\lambda \end{pmatrix} \tag{3.23}$$

The determinant, which represents the characteristic polynomial of the original matrix, is then computed. We find the determinant by applying Laplace's theorem. Finally, we multiply the diagonals of the matrices. The calculations can be finalized, comparable terms combined, and the distinctive base-three polynomial fixed.

$$\begin{aligned} |A_\lambda| &= -\lambda \begin{vmatrix} -\lambda & 1 \\ 1 & -\lambda \end{vmatrix} - 1 \begin{vmatrix} 1 & 1 \\ 1 & -\lambda \end{vmatrix} + 1 \begin{vmatrix} 1 & -\lambda \\ 1 & 1 \end{vmatrix} \\ &= -\lambda\left(\lambda^2 - 1\right) - 1\left(-\lambda - 1\right) + 1\left(1 + \lambda\right) \\ &= -\lambda^3 + \lambda + \lambda + 1 + 1 + \lambda \\ &= -\lambda^3 + 3\lambda + 2 \end{aligned} \tag{3.24}$$

The polynomial is set to zero (first we change its sign). The first eigenvalue is the value that makes the polynomial equal to zero, in this case -1.

$$\lambda^3 - 3\lambda - 2 = 0 \qquad \lambda = -1 \tag{3.25}$$

Ruffini's rule is used to reduce the polynomial's degree in order to find the remaining eigenvalues.

	1	0	-3	-2
-1		-1	1	+2
	1	-1	-2	0

3.26

↓

$$Q(\lambda)=\lambda^2-\lambda-2\rightarrow\lambda=\frac{1\pm\sqrt{1+8}}{2}=\rightarrow\lambda=-1\,;\,\lambda=2 \quad 3.27$$

We observe that two eigenvalues in this instance have the same value, namely -1. We only keep the highest eigenvalue ($\lambda = 2$) out of the three that were found, as indicated in the initial formula.

Since our adjacency matrix is composed of three nodes, A_λ is equal to the vectors x, y, and z, which must equal the null column vector to determine the eigenvector. It needs eigenvectors that are not zero. Then we replace the eigenvalue 2 with the diagonal A_λ. The matrix-vector product is given by the multiplication of each row of the matrix with the vector x, y and z. The eigenvector cannot be zero, so three system equations must be solved with this in mind.

$$\begin{aligned} A_\lambda &= \begin{pmatrix} x \\ y \\ z \end{pmatrix} = \begin{pmatrix} 0 \\ 0 \\ 0 \end{pmatrix} = \begin{pmatrix} -2 & 1 & 1 \\ 1 & -2 & 1 \\ 1 & 1 & -2 \end{pmatrix} \bullet \begin{pmatrix} x \\ y \\ z \end{pmatrix} \\ &= \begin{pmatrix} 0 \\ 0 \\ 0 \end{pmatrix} \rightarrow \begin{cases} -2x+y+z=0 \\ x-2y+z=0 \\ x+y-2z=0 \end{cases} \rightarrow \begin{pmatrix} 1 \\ 1 \\ 1 \end{pmatrix} \end{aligned} \tag{3.28}$$

The eigenvector centrality for each node of the initial adjacency matrix is determined. In this example, since every node is connected to every other node, all nodes are assumed to have an eigenvector centrality equal to 1. In the following session, we will look at a derivation of eigenvector centrality known as the Katz centrality.

3.5.4 *Katz Centrality*

Katz centrality, developed by Katz (1953), is a measure used to determine the relative importance of a specific node within a network. Unlike other measures of centrality, which only consider the shortest path (the geodesic) between two nodes, Katz centrality considers all walks between nodes. It calculates the relative influence of a node within a network by considering all paths that originate from the node under consideration (Zinilli & Cerulli, 2023). It is assumed that nodes closer to the node under consideration have more influence than those further away. Consequently, Katz centrality weights each path's contribution by an attenuation factor α raised to the power of the path's length. Thus, the Katz centrality is defined as:

$$C_K\left(i\right)=\sum_{k=1}^{+\infty}\sum_{j=1}^{n}\alpha^k\left(A^k\right)ji \tag{3.29}$$

α is nothing more than a weight that prioritizes shorter pathways. The attenuation factor α should be smaller than the reciprocal of the absolute value of the largest eigenvalue of A. A^k includes elements that represent the number of pathways of length k connecting nodes i and j. Elements a_{ij} of A are vectors that have the value 1 if a node i is linked to a node j and 0 otherwise. After that, the following equation is used to calculate Katz centrality

$$C_K=\left(\left(I-\alpha A^T\right)^{-1}-I\right)I \tag{3.30}$$

where I denotes the identity matrix. A^T represents the transposed matrix of A, while $(I\text{-}\alpha A^T)^{-1}$ represents the inverse of $(I\text{-}\alpha A^T)$. In the case of oriented networks, we have to consider the number of outbound and inbound edges from the node i, and the Katz centrality is:

$$\begin{aligned} C_k^{out}&=\left(\left(I-\alpha A^T\right)^{-1}-I\right)\\ C_k^{in}&=\left({}^T\left(I-\alpha A^T\right)^{-1}-I\right)\end{aligned} \tag{3.31}$$

The attenuation factor, typically denoted as α, controls the relative influence of nodes that are farther away in the network. In the Katz centrality measure, a node's influence is determined by the sum of the contributions of all other nodes that can reach it through a directed path, with each path's contribution attenuated by α to the power of its length. This effectively applies an exponential decay based on the path length. The choice of α depends on the characteristics of the network under analysis. In general, smaller values of α (closer to 0) give more importance to nodes that are closer to the node of interest by significantly attenuating the influence of distant nodes. Conversely, larger values of α (closer to 1) reduce this attenuation, which may allow distant nodes to contribute more significantly to the centrality, but care must be taken as excessively large values can make the centrality measure unstable. PageRank is discussed in the next paragraph; it shares similarities with Katz centrality. Both are measures of node importance in a network that take into account not just the node's direct connections, but also the importance of the connected nodes. However, in PageRank, the importance of a node (viewed as a finite resource) is distributed among its outgoing edges. In contrast, Katz centrality counts all paths from a node, with the contribution of each path attenuated by its length.

3.5.5 PageRank Centrality

PageRank is a measure of global centrality that uses the entire network to assess the significance of a single node. Similar to eigenvector centrality, PageRank assesses a node's relevance based on the importance of the nodes that link to it. It calculates the centrality of a node based on the importance of its neighbors. Thus, it is an iterative process that estimates a node's importance in the network using global information. In the first iteration, each node is equally relevant:

$$PR_i^0 = \frac{1}{n}, \forall i \tag{3.32}$$

where n is the number of vertices/pages in the graph. Each node distributes its centrality equally to the nodes it links to (outgoing). The PageRank of a node is calculated by adding the PageRank fractions of the vertices that have edges to it from the previous phase (PageRank/rank in output):

$$PR_i = \sum_{j=1}^{n} a_{ji} \frac{PR_j}{d^{out}}, \forall t \in \{1,k\} \tag{3.33}$$

where d_j^{out} is the number of hyperlinks on the node/page j. The first iteration is then:

$$PR_i^1 = \frac{1}{n} \sum_{j=1}^{n} a_{ji} \frac{1}{d_j^{out}} \tag{3.34}$$

The next steps provide an example of computing a simplified PageRank algorithm. We have a graph composed of 5 nodes (Fig. 3.7).

Iteration 1 initializes all nodes/pages to have rank 1/5. In the iteration 2, the node A has one edge from C, and C has four outbound edges, so we take the rank of C from iteration 1 and divide it by 4, which results in rank (1/5)/4 = 1/20 for the node A. This process is used for all vertices. Table 3.1 is an example of PageRank computation based on the graph in Fig. 3.7.

$$\mathrm{PR_A} = \frac{(1/5)}{4} = \frac{1}{20}$$

$$\mathrm{PR_B} = \frac{(1/5)+(1/5)}{4} = \frac{5}{20}$$

$$\mathrm{PR_C} = \frac{(1/5)}{2} = \frac{1}{20}$$

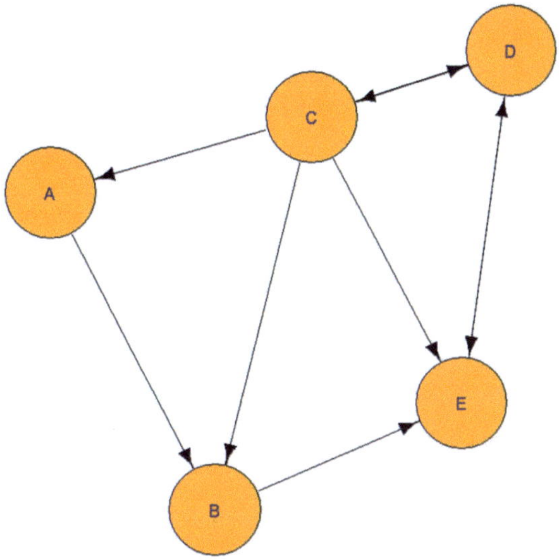

Fig. 3.7 Graph example

Table 3.1 Example of PageRank computation (first step)

Nodes	Iteration 1	Iteration 2
A	1/5	1/20
B	1/5	5/20
C	1/5	1/10
D	1/5	5/20
E	1/5	7/20

$$PR_D = \frac{(1/5)}{4} + \left(\frac{1}{5}\right) = \frac{5}{20}$$

$$PR_E = \frac{1}{5} + \frac{(1/5)}{4} + \frac{(1/5)}{2} = \frac{7}{20}$$

The algorithms repeat the same process for additional steps after the second iteration. Because PageRank can fluctuate, the algorithm iterates more times. Multiple iterations are typically required to accurately compute the PageRank. Iterations stop when the PageRank values converge, namely when the PageRank at iteration $t + 1$ differs from iteration t by a very low value tending to zero. In Table 3.2, we stopped at iteration 3 with a given final rank.

The simple update rule can result in PageRank building up and becoming concentrated in specific areas of some graphs. For instance, some nodes/pages do not get a PageRank score when they have no incoming edges. In this situation, some variations of PageRank employ a so-called damping factor (a constant). The damping factor helps prevent sinks (nodes/pages with no outbound edges) and ensures each node/page receives a PageRank score. When creating the ranking module for the Google prototype in 1998, Brin and Page provided the initial formula for

Table 3.2 Example of PageRank computation (second step)

Nodes	Iteration 1	Iteration 2	Iteration 3	Final rank
A	1/5	1/20	1/40	5
B	1/5	5/20	3/40	4
C	1/5	1/10	5/40	3
D	1/5	5/20	15/40	2
E	1/5	7/20	16/40	1

calculating PageRank (Brin & Page, 1998). The World Wide Web may be shown as a huge oriented network, with nodes representing web sites and arcs representing connections connecting them. For instance, during an Internet search, the process of discovering results ensures that websites fitting our search parameters are displayed.

3.6 Community Detection

Community detection analysis is currently receiving a lot of attention. Finding groups of nodes that are, in some way, more similar to each other than to other nodes is the fundamental goal of community detection in Network Science. There are various types of communities that can be identified in a network, such as overlapping communities (where nodes can belong to multiple communities) and hierarchical communities (where communities can be nested within larger communities). Communities can be described as subgroups of vertices that have a higher density of internal connections than external edges (Newman, 2018). Subgroups and their cohesion can be studied to identify closed groups of individuals who can act as a barrier to the spread of knowledge and innovation because they are highly interconnected and thus resistant to outside influences.

In the next paragraphs, we will deal with the most used community measures, such as the clustering coefficient and Rich-Club concept.

3.6.1 *Clustering Coefficient*

Clustering coefficient, also known as transitivity, is a feature of real networks that increases the probability that two unconnected vertices who share an edge with a third vertex will connect.

Transitivity consequently suggests that there are many triangles present in a generic network *G*. The clustering coefficient *C*, a statistic developed by Watts and Strogatz (1998), can be used to quantify the number of triangles in a graph. First, we introduce a statistic $C_{local}(i)$ (the local clustering coefficient of vertex *i*) which describes how likely it is that $a_{jm} = 1$ for vertex *i*'s two neighbors *j* and *m*. The average of the $C_{local}(i)$ extended to all vertices gives the network's global clustering coefficient.

The clustering coefficient reflects the degree to which a vertex's neighbors are related to one another. This measure is frequently used to understand how vertices in a network tend to cluster together, as well as whether a graph has a small world, random, or scale-free property (in the next chapter we will address this issue).

3.6.1.1 Local Clustering Coefficient

There are different definitions of local clustering coefficient. Here, we refer to the definition of Watts and Strogatz (1998). Assuming there is a node i of degree k, and let b_i to be the number of edges that exist between the k neighbors of i. The local clustering coefficient, $C_{local}(i)$, is the proportion between the actual number of edges among the neighbors, b_i, and the maximum possible value of edges among the neighbors of i. Maximum possible value calculated through the classic binomial coefficient notion $n!/[k!(n\text{-}k)!]$. The local clustering coefficient for undirected graphs is then defined by:

$$C_{local}(i) = \frac{b_i}{\frac{n!}{k!(n-k)!}} \qquad \text{with } 0 \leq C_{local}(i) \leq 1 \tag{3.35}$$

Thus, the average likelihood that two neighbors of node i are also connected is measured by this local clustering coefficient. The global clustering coefficient of a graph is defined as the average value of $C_{local}(i)$ across all nodes in the network:

$$C_{global}(i) = \frac{1}{g}\sum_{i=1}^{g} C_{local}(i) \tag{3.36}$$

where g is the total number of nodes in the observed graph. Referring to the next graph example with 5 nodes and 6 edges (Fig. 3.8), we compute the local and global clustering coefficients by hand.

$$C_{local}(A) = \frac{1}{\frac{n!}{k!(n-k)!}} = \frac{1}{\frac{3!}{2!(3-2)!}} = \frac{1}{\frac{6}{2}} = 0.\overline{3} \tag{3.37}$$

The local clustering coefficient for the node A is 0.3. The higher the connectivity of neighbors of a specific node in the graph, the closer the local clustering coefficient is to one.

Fig. 3.8 Graph example

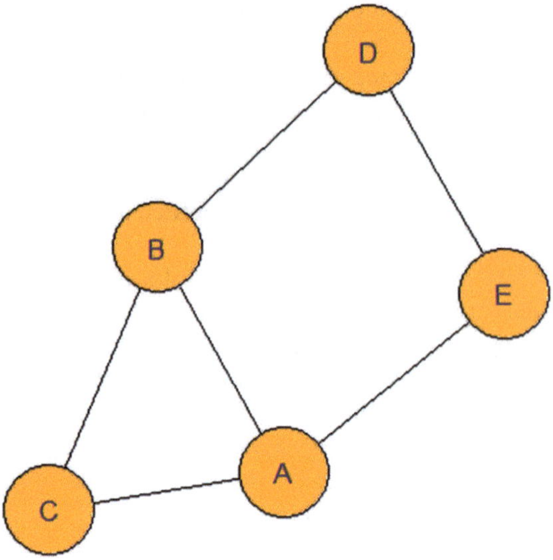

3.6.2 Rich-Club

While researching the internet architecture, Zhou and Mondragón (2004) first proposed the Rich-Club theory in 2004. A Rich-Club denotes the presence of a collection of nodes that have a large number of connections between them and are generally strongly linked to one another, forming a close-knit group (club). Similar to other coefficients and metrics we are covering in this book, the Rich-Club coefficient can be determined using a variety of procedures. Here, we mainly provide a general overview of the Rich-Club coefficient calculation; however, if you are interested in learning more, check Zhou and Mondragón (2004), Colizza et al. (2006), and Jiang and Zhou (2008).

For each degree k, the Rich-Club coefficient is defined as the ratio of the number of observed edges to the number of possible edges for nodes with degree larger than k:

$$\phi(k) = \frac{2E_k}{N_k(N_k - 1)} \tag{3.38}$$

with N_k is the number of nodes with degree larger than k, and Ek is the number of edges among those vertices. Given that the $N > k$ nodes are grouped into a clique, it follows that $\phi(k) = 1$. When compared to its expectation over a set of rewired networks with the same degree sequence as the original one, the Rich-Club coefficient is known as the *(k)norm*. While a network is said to exhibit Rich-Club ordering when $\phi(k)norm > 1$. When investigating Rich-Club ordering, the Rich-Club coefficient should always be compared to a null model in order to determine its significance. This makes the normalization of the Rich-Club coefficient a crucial step to be taken in order to account for the significance of this measure. The normalized Rich-Club coefficient (Colizza et al., 2006) is often defined as follows:

$$\phi(k)_{norm} = \frac{\phi(k)}{\phi_{rand}(k)} \tag{3.39}$$

where $\phi_{rand}(k)$is the average Rich-Club coefficient estimated with the same degree distribution as the original network using an appropriate null model. Typically, the average Rich-Club coefficient is estimated across 1000 rewired networks (De Domenico & Arenas, 2017). When compared to the null scenario, a normalized Rich-Club coefficient larger than 1 indicates the existence of the Rich-Club phenomenon. Iterative rewiring based on several steps is used to construct the null model. The procedure provides for selecting two random edges at each step and exchanging one of the first edge's endpoints for one of the second edge's endpoints, whether the edges generated are unique and do not already exist.

3.7 Homophily and Similarity

The term "homophily" basically means the creation of edges between nodes according to how similar they are. Homophily is a common structural regularity in real-world networks: it is the observed tendency of nodes with similar features to establish relationships with one another (Jackson, 2010). As argued by Resce et al. (2022), homophily, or similarity in terms of specific node features, has a significant impact on the formation of relationships.

Different measures can be used to determine homophily; in this article, we analyze three of the most important ones that are frequently found in the main statistical software and are used in the social sciences: the E-I Index, Cosine similarity, and Jaccard coefficient.

3.7.1 E-I Homophily Index

As a measure of homophily, the E-I index describes actors' propensity to interact with similar actors (Burt, 1991). The E-I index is an indicator of in- and out-group preference (Krackhardt & Stern, 1988). The E-I index is computed by subtracting the number of in-group ties from the number of out-group ties and dividing the result by the total number of ties (Everett & Borgatti, 2012). The E-I index specifies how many ties a graph (two or more actors who interact with one another) has to its own members vs how many connections it has to those outside the group (Krackhardt & Stern, 1988). The basic equation is

$$E-I\ Index = \frac{E-I}{E+I} \tag{3.40}$$

where I (internal) is the number of edges within a specific group and E is the number of edges to the external group. It can range between −1 and 1, where 1 denotes complete heterophily and − 1 denotes complete homophily. Since it captures any tie between members of the same group, the direction of edges is not considered (i.e., either an out-edge or an in-edge constitutes an edge between two nodes). The E-I index is an appealing metric of homophily since it is not affected by network density. Because it does not take into account the underlying population, it can be considered as a relative measure of homophily (Everett & Borgatti, 2012).

Even though the E-I index is most typically employed in unweighted networks, some scholars also used it in weighted networks (De Andrade & Rêgo, 2018). In this case, E is the total weight of edges connecting nodes in different partitions, while I is the total weight of edges within the same partition.

For more information, De Andrade and Rêgo (2018, 2022) provide further details on the usage and computation of the E-I index in weighted networks.

The following example (Table 3.3) represents a network of 15 people from four different countries (Italy, France, Germany and Austria). The values within each cell indicate the links between people with respect to nationality. Using the previous equation, we will determine the E-I index for this network.

The values on the diagonal of the table show the number of relationships between persons from the same country (9), while the values off the diagonal represent links between individuals from different nationalities (6). Under the equation applied to the example's values:

$$E - I\,Index = \frac{6-9}{15} = -0.2 \tag{3.41}$$

The E-I index is −0.2, suggesting a low level of homophily. Generally, after calculating the E-I index of a network partition, a permutation test is used to determine its significance. Question: is the observed E-I index higher or lower than expected? If the links in a network are randomly assigned, we would expect the proportion of links between similar people (in this case, based on nationality) to reflect the overall composition of the network. In our example, the network has 15 connections. How many relationships would there be between people with the same nationality if they were assigned at random? A permutation test is used to determine whether the observed E-I index is considerably higher or lower than expected values. Typically,

Table 3.3 Example of a network for the application of the E-I index

	Total	Italy	France	Germany	Austria
Italy	6	2	2	1	1
France	3	0	1	1	1
Germany	5	0	0	5	0
Austria	1	0	0	0	1
Total	15	2	3	7	3

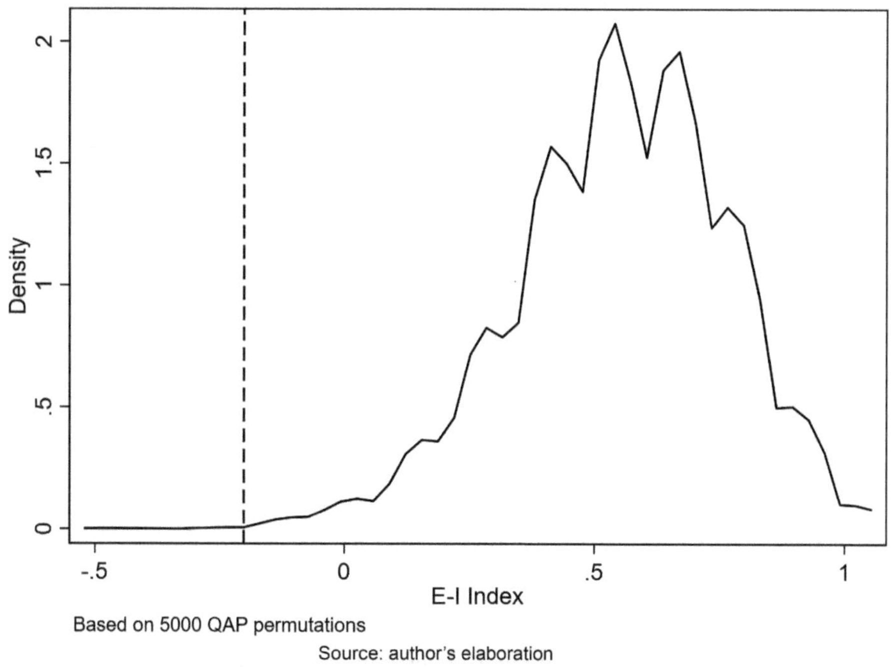

Fig. 3.9 Permutation test distribution based on the example data in Table 3.3

the number of permutations is more than or equal to 5000. Permutation tests are non-parametric tests (no assumption about the underlying distribution) with few assumptions. The main assumption required is the exchangeability hypothesis (permutation-invariant), that is, the joint distribution of the observations is invariant under permutations. In a nutshell, the nodes are exchanged without affecting the probability distribution (we shall return to this subject in Chap. 5). This kind of test is a valuable approach for establishing statistical significance when we do not know much about the data producing process, namely the target population. To determine whether the observed E-I index is more or less probable to be from the reference distribution, an empirical p-value is calculated.

The permutation test resulting from the information in Table 3.3 is shown in Fig. 3.9.

Compared to the observed, all the generated networks have a higher E-I index value. Thus, the empirical p-value will surely tend to zero. The dotted vertical line is our observed E-I index (−0.2). The empirical p-value is 0.0012 since 99.88% of our permutation distribution is above the dotted line.

3.7.2 Cosine Similarity

The cosine, which provides us with a measure of similarity (in terms of angle) between two vectors, might be used to express the degree of similarity between the nodes in a graph. Therefore, in geometry, we have:

$$x \cdot y = |x||y|\cos\theta \tag{3.42}$$

where θ is the angle between the vector x and the vector y, as well as $|x| = \sqrt{x^2}$ and $|y| = \sqrt{y^2}$.

Therefore, considering the i-th and j-th rows (and columns) of the adjacency matrix as two vectors, we can define the cosine of the angle between the two as our measure of similarity σ_{ij}:

$$\sigma_{ij} = \cos\theta = \frac{x \cdot y}{|x||y|} = \frac{\sum_k A_{ik} A_{kj}}{\sqrt{\sum_k A_{ik}^2}\sqrt{\sum_k A_{kj}^2}} \tag{3.43}$$

In the case of an unweighted graph, we have $A_{ij}^2 = A_{ij}$ and we can get the denominator by adding $\sum_k A_{ik} = k_i$, namely the degree of node. Consequently, our formula is summarized as follows:

$$\sigma_{ij} = \frac{\sum_k A_{ik} A_{kj}}{\sqrt{k_i k_j}} = \frac{\mathrm{n}_{\mathrm{ij}}}{\sqrt{k_i k_j}} \tag{3.44}$$

where n_{ij} represents the number of nodes in common between the nodes i and j (to which they are both connected). We calculate the similarity between A and B for the graph in Fig. 3.10:

$$\sigma_{AB} = \frac{n_{AB}}{\sqrt{k_A k_B}} = \frac{1}{\sqrt{2^* 2}} = 0.5 \tag{3.45}$$

The numerator in Eq. 3.45 represents the number of shared neighbors between nodes A and B, with node C being the only common neighbor. The denominator includes k, which denotes the degrees of nodes A and B, respectively.

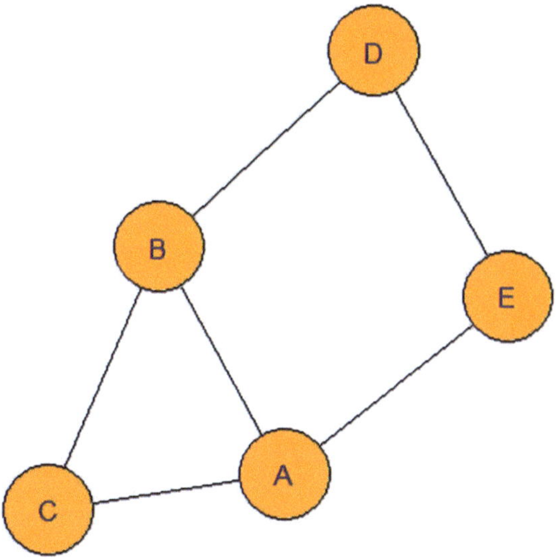

Fig. 3.10 Graph example

3.8 Structural Holes

Structural holes describe the situation in which there are gaps, or holes, between different clusters of network nodes. Burt (2000, p. 65) refers to structural holes as: "*…a relationship of no redundancy between two contacts*" and "*As a result of the hole between them, the two contacts provide network benefits that are in some degree additive rather than overlapping*". Most real-world networks tend to be characterized by different structures, e.g., dense clusters of strong connections (network closure), clusters of weak connections, and structural gaps. Structural holes are relevant because they represent opportunities for the flow of diverse information. Because the broker can acquire diverse, non-redundant information from the two disconnected groups, a person who serves as a bridge between two or more closely linked groups of individuals may gain considerable competitive advantages.

The structural hole theory was driven by Granovetter's weak tie theory (Granovetter, 1973). Strong edges, following the Mark Granovetter's theory, lead to homophily. According to the weak tie theory, the overlap of two nodes in their relationship network develops as the strength of their relationship grows. For instance, a weak tie serves as channel for the spread of innovative ideas across different groups. A broker is also in a vulnerable position since links with various groups can be weak and time consuming to maintain. The following figure (Fig. 3.11) describes the weak and strong connections that exist among the nodes beyond the structural holes. The dotted lines indicate weak connections, whereas the solid lines indicate strong connections; the two green curly brackets in Figure 3.11 represent structural holes.

The concept of structural holes involves examining the structure of a node's local connections, considering aspects such as redundancy, effective size, and constraint,

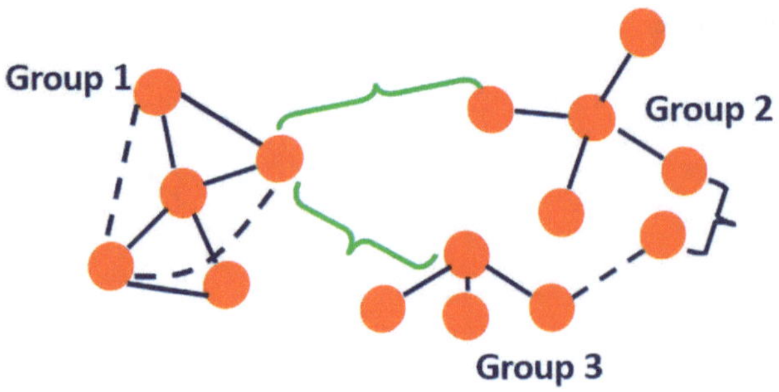

Fig. 3.11 Example of a graph with structural holes, as well as weak and strong connections

as defined by Burt (1992). About the concept of network redundancy, Burt wants to know how much contact j and other contacts of node i overlap. In terms of relationship investment, when a node j is strongly connected to another node q, redundancy could be viewed as devoting time and energy to a relationship that does not provide new or unique connections or information

$$R = p_{iq} m_{jq} \quad \text{with}\, q \neq i, j \tag{3.46}$$

where p_{iq} is the ratio of i's energy used to interact with q, and m_{jq} is calculated as j's interaction with q divided by j's strongest relationship with anyone. By adding up this product across all nodes q, the redundancy in a network is determined. To capture the redundancy in a network, many metrics are applied. For instance, Borgatti (1997) proposed an alternative method for calculating the effective redundancy of an unweighted network:

$$R = \frac{2t}{n} \tag{3.47}$$

where t is the total number of edges in the egocentric network and n is the total number of nodes in the network.

The effective network size of generic node i is defined as the sum of j's nonredundant contacts:

$$S_i = \sum_j \left[1 - \sum_q p_{iq} m_{jq} \right] \tag{3.48}$$

The larger the effective size, the greater the distance between each node and the other main contacts. This indicator ranges from one (a single network connection) to the total number of edges N (each contact is unique).

Another common measure for structural holes is the constraint. Network constraint is the sum of the constraints of each connection c_{ij}:

$$c_{ij} = \left(p_{ij} + \sum_{q} p_{iq} m_{jq} \right)^2 \tag{3.49}$$

Nodes with lower constraint values are less constrained by their direct neighbors, allowing them to access new information outside of a tightly knit group of neighbors. Higher constraint value nodes are constrained by their direct neighbors.

3.9 Entropy

Entropy is a concept used to measure the degree of disorder or uncertainty in a system. The entropy of a system is a measure of the amount of randomness or unpredictability in the system. In mathematical terms, entropy is often defined as a negative logarithmic measure of the probability of an event. The entropy of a discrete random variable X with possible values $x_1, x_2, x_3, \ldots, x_n$ and corresponding probabilities $p_1, p_2, p_3 \ldots, p_n$ can be represented as:

$$H(X) = -\sum_{x \in X} p_i \log(p_i) \tag{3.50}$$

log denotes the logarithm to base 2. The logarithm base 2 ensures that the entropy is measured in bits, which is a unit of information. In this formulation, entropy is a scalar value that ranges from 0 to *log(n)*, where 0 indicates a completely predictable system (i.e., all the probabilities are 1 for a single value) and *log(n)* indicates a completely random system (i.e., all the probabilities are equal). Therefore, the entropy is always non-negative and reaches its maximum when all states are equally likely, meaning that the system exhibits maximum uncertainty or disorder. As the probabilities of each state become more skewed, the system becomes less uncertain and the entropy decreases.

Here is a simple example of how to calculate entropy.

Suppose we have a system with two possible states, *A* and *B*, and the probability of being in state *A* is 0.7, and the probability of being in state *B* is 0.3. The entropy of this system can be calculated using the formula:

$$H = -p_1 \times \log_2(p_1) - p_2 \times \log_2(p_2) \tag{3.51}$$

where *p1* and *p2* are the probabilities of being in states *A* and *B*, respectively. Plugging in the values, we get:

$$H = -0.7 \times \log_2(0.7) - 0.3 \times \log_2(0.3) = 0.881 \text{ bits} \tag{3.52}$$

Thus, the entropy of this system is 0.881 bits. This means that the system is relatively uncertain, as there is a 70% chance of being in one state and a 30% chance of being in the other state.

Though this is a simplified example, the idea of entropy is applicable to many more settings and systems. Entropy finds applications in the social sciences, where it is used to quantify uncertainty in various systems. These include the diffusion of ideas within a group, the allocation of resources within a society, and the distribution of power within a political structure. Entropy can be used to evaluate the degree of stability or change in a system over time and to compare several systems.

3.9.1 Stata Code and Output

```
// We use the Zachary's Karate Club graph
. clear all
. nwset, nwclear

// Uploading a network from other sources: UCINET
. nwimport http://vlado.fmf.uni-lj.si/pub/networks/data/ucinet/zachary.dat, type (ucinet)

// Getting the degree for each node. In the case of a direct network we have to add: in(generate(indeg)) out(generate(outdeg))
. nwdegree ZACHC, generate(degree) isolates standardize

// Getting the neighbors of node 1
. nwneighbor ZACHC, ego (1)

// Computing the average degree of neighbors of the Node 1
. nwcontext ZACHC, attribute(degree) stat(mean) generate(wmean)

// Getting the farness, nearness and closeness centrality for each node.
. nwcloseness ZACHC , generate(farness nearness closeness)
```

$$Farness(i) = \frac{1}{N-1}\sum_{j=1}^{N} l_{ij}$$

$$Nearness(a) = 1 / Farness(a)$$

$$\text{Closeness}(i) = Nearness(i) * (\text{nodes} - 1)$$

```
// Getting the betweenness centrality for each node.
. nwbetween klas12b_wave1

// Getting the eigenvector for each node.
. nwevcent ZACHC, gen(eigen) nosym

// Katz centrality calculation for each node. Alpha is the attenuation factor.
. nwkatz ZACHC, alpha(0,5)

// Getting the local clustering coefficient for each node, as well as the overall and average clustering coefficient.
. nwclustering ZACHC

// Obtain the homophily (E-I index) for the degree centrality.
. nwtabulate ZACHC degree

// Correlate nodes of the network. The node i's tie vector and node j's tie vector are taken and correlated.
. nwcorrelate ZACHC, attribute(degree) permutations(500)

// Calculate Burt's structural hole measures: effective size, efficiency, and constraint
. nnwburt ZACHC

// Generate Jaccard matrix distance (Jaccard Coefficient)
. nwsimilar ZACHC, type(jaccard)
```

3.9.2 R Code and Output

```
# Downloads the Zachary's Karate Club graph
> install.packages("CINNA")
> library(CINNA)
> data("zachary")
> print(zachary)

# Loading package igraph and network
> library(igraph)
> library(network)

# Now let us duplicate the Zachary dataset and give it the name g
> g<-zachary

# Identify all vertex 1 neighbors, regardless of their orientation
> neighbors(g, '1', mode = c('all'))

# Obtaining any nodes, as well as their intersection, that take an edge from vertex 1 and send it to vertex 2
> n1 <- neighbors(g, '1', mode = c('out'))
> n2 <- neighbors(g, '2', mode = c('in'))
> intersection(n1, n2)

# Calculate the out-degree of each node (for the in-degree the mode is "in")
> g.Outd <- degree(g, mode = c("out"))

# View the out-degree summary
> table(g.Outd)

# Find the node that has the maximum out-degree
> which.max(g.outd)

# Calculate the betweenness centrality of each node
> gbet <- betweenness(g, directed = TRUE)

# Create a plot where the node size is based on the betweenness score
> plot(g, vertex.label = NA, edge.color = 'black', vertex.size = sqrt(gbet)+7, edge.arrow.size = 0.08, layout = layout_nicely(g))
```

Output

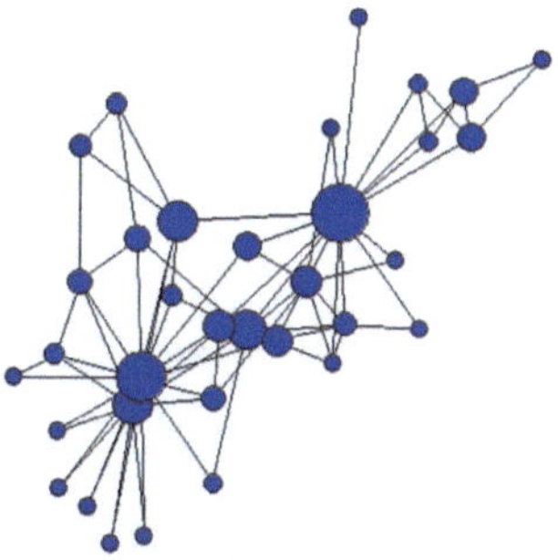

```
# Retrieve the strength or weighted degree of a node
> strength <- strength(g)

# Get the closeness centrality
> closeness <- closeness((g), normalized=TRUE)

# Get the betweenness centrality
> betweenness <- betweenness(g, normalized=TRUE)

# Get the eigenvector centrality
> igencent <- eigen_centrality(g)
> g.eigencent <- igencent$vector

# The clustering coefficient, commonly known as the transitivity measure. Specify the type of transitivity to calculate (e.g., global, local, weighted)
> transitivity<-transitivity(g, type = c("global"),weights = F )

# Calculate the density of the graph
> density<- edge_density(g)
```

```
# Order the number of nodes and edges of a graph
> Nodes<-gorder(g)
> Edges<-gsize(g)

# Getting some centralization measures from the centrality scores of the nodes
> g_cent<-graph_from_adjacency_matrix(as.matrix(g), mode="undirected", weighted=TRUE)
> g_cent <- centr_degree(g)$centralization
> g_cent[2] <- centr_clo(g)$centralization
> g_cent[3] <- centr_betw(g)$centralization
> g_cent[4] <- centr_eigen(g)$centralization
> g_cent[5] <- transitivity
> g_cent[6] <- density
> g_cent[7] <- Nodes
> g_cent[8] <- Edges

# Combine the multi-dimensional arrays, then display the results in a table
> tab.cent <- rbind(g= g_cent)
> dimnames(tab.cent)[[2]] <- c("Degree", "Closeness", "Betweenness", "Eigenvector", "Transitivity", "Density", "Nodes",
"Edges")
> tab.cent

# Calculate the assortativity degree of the graph
> assortativity <- assortativity.degree(g, directed = FALSE)

# Calculate the assortativity for the degree centrality. Interconnection of highly connected nodes with other highly connected nodes
> degree_assortativity <- assortativity.degree(g)

# Calculate the assortativity for a specific node attribute. Below the interconnection between those of similar betweenness centrality
> bet_assortativity<-assortativity(g, sample(gbet))

# Determine the reciprocity of the graph
> reciprocity(g)

# Computing the Burt's constraint for each vertex
> constraint(g, nodes = V(g), weights = NULL)

# Get the modularity maximization via the Greedy algorithm and determine the sizes of each community. This algorithm is agglomerative: at each stage, two groups are combined. The merger is determined by maximizing modularity. This is a quick algorithm, though it has the drawback of being greedy. As a result, it may not generate the best overall community partitioning, but it is useful and accurate.
> Mod <- fastgreedy.community(g)
> sizes(Mod)

# Determine which individuals belong to which community
> membership(Mod)
> plot(Mod, g, edge.color = 'black')
```

Output

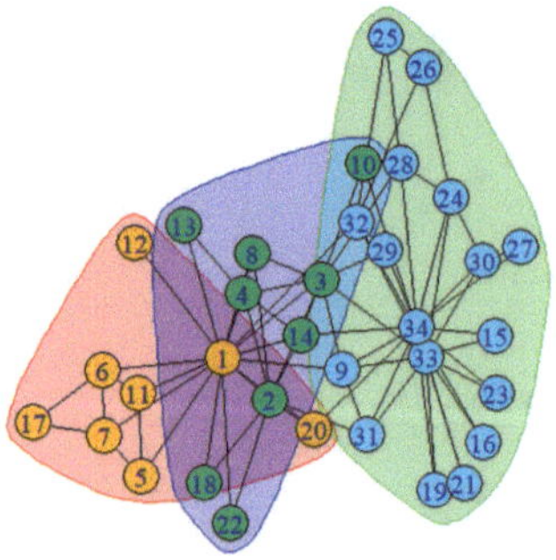

```
# Rich-Club function. The 'normalized' parameter is a boolean value indicating whether the Rich-Club
coefficient should be normalized by the maximum possible value

> richClub <- function(g, normalized=FALSE) {
  degree <- colSums(g)
  k_min <- min(degree)
  k_max <- max(degree)
> rich_club_coef <- vector(mode="numeric", length=(k_max-k_min+1))
> for (k in k_min:k_max) {
   nodes <- which(degree >= k)
   subgraph <- g[nodes,nodes]
   num_edges <- sum(subgraph) / 2
   num_possible_edges <- sum(degree[nodes]) - k * length(nodes) / 2
   rich_club_coef[k-k_min+1] <- num_edges / num_possible_edges
 }

 # Normalisation
> if (normalized) {
     max_coef <- max(rich_club_coef)
     rich_club_coef <- rich_club_coef / max_coef
  }

 # Returns the vector of coefficients Rich-Club
 return(rich_club_coef)
}

# Create a random graph
> set.seed(123)
> g <- erdos.renyi.game(20, 0.2)

# Calculate the degree of each node
> deg <- degree(g)

# Sort the nodes according to their degree
> sorted_nodes <- order(deg, decreasing = TRUE)

# Initialise a vector for the Rich-Club coefficients
> rc_coef <- rep(0, length(sorted_nodes))

# For each degree, calculate the number of arcs within the subgraph of the Rich-Club
> for (k in 1:length(sorted_nodes)) {
 rc_nodes <- sorted_nodes[1:k]
 rc_subgraph <- induced_subgraph(g, rc_nodes)
 rc_edges <- ecount(rc_subgraph)

 # Calculate the normalized Rich-Club coefficient
 rc_coef[k] <- 2*rc_edges/(k*(k-1))
}

# Set colour palette
> library(RColorBrewer)
> colors <- brewer.pal(2, "Set1")

# Set background color
> bg_color <- "#F5F5F5"

# Plot of Rich-Club coefficients
> plot(seq_along(sorted_nodes), rc_coef,
    type = "l", lwd = 2, col = "red",
    xlab = "Degree", ylab = "RichClub Coefficient",
    main = "Rich-Club Coefficient Plot",
    cex.lab = 1.2, cex.main = 1.5)

# Add legend
> legend("topright", legend = "Rich-Club Coefficient",
     lty = 1, lwd = 2, col = "red")
```

Output

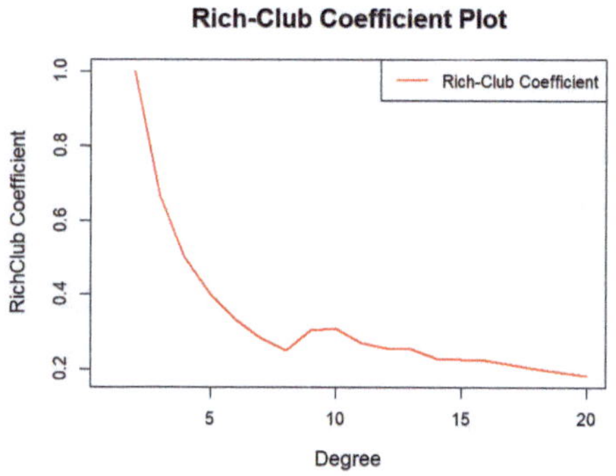

3.9.3 Python Code and Output

```
# Downloads the necessary libraries
>>> import matplotlib.pyplot as plt
>>> import networkx as nx
>>> import pandas as pd
>>> import numpy as np

# We use the Zachary's Karate Club graph
>>> g = nx.karate_club_graph()
>>> g1= nx.karate_club_graph()
>>> nx.draw(g)

# Indicate if the graphs g and g1 are isomorphic [True or False]
>>> nx.is_isomorphic(g, g1)

# Average degree calculation
>>> N, K = g.order(), g.size()
>>> avg_deg = float(K) / N
>>> print ("Nodes: ", N)
>>> print ("Edges: ", K)
>>> print ("Average degree: ", avg_deg)

# Calculate the degree centrality of the graph g. For directed graphs, use nx.in degree centrality and nx.out degree centrality for in-degree and out-degree, respectively
>>> deg_cen= nx.degree_centrality(g)
>>> print(deg_cen)
>>> pos_g = nx.spring_layout(g ,iterations = 1000)
>>> nsize = np.array ([1-v for v in deg_cen.values()])
>>> nsize = 10**(nsize+1)
>>> nodes = nx.draw_networkx_nodes (g , pos = pos_g , node_color=nsize, cmap=plt.cm.coolwarm,node_size = nsize)
>>> edges = nx.draw_networkx_edges (g , pos = pos_g , alpha = .1)
>>> plt.axis('off')
```

Output

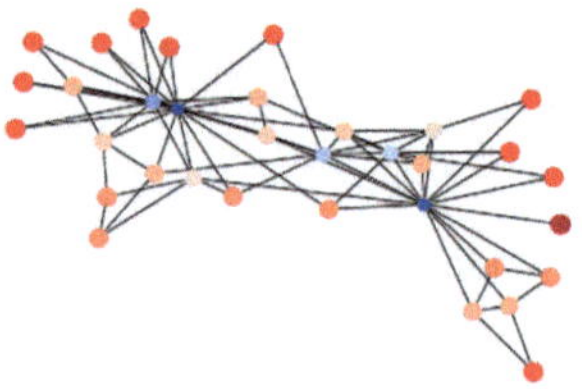

```
# Visualizing the distribution of the degree centrality through the degree-rank plot and the degree histogram
>>> degree_sequence = sorted((d for n, d in g.degree()), reverse=True)
>>> dmax = max(degree_sequence)
>>> fig = plt.figure("Degree of Zachary graph", figsize=(8, 8))
# Create a gridspec for adding subplots of different sizes
>>> axgrid = fig.add_gridspec(5, 4)

>>> ax0 = fig.add_subplot(axgrid[0:3, :])
>>> Gcc = g.subgraph(sorted(nx.connected_components(g), key=len, reverse=True)[0])
>>> pos = nx.spring_layout(Gcc, seed=10396953)
>>> nx.draw_networkx_nodes(Gcc, pos, ax=ax0, node_size=20)
>>> nx.draw_networkx_edges(Gcc, pos, ax=ax0, alpha=0.4)
>>> ax0.set_title("Connected components of G sub-graph")
>>> ax0.set_axis_off()

>>> ax1 = fig.add_subplot(axgrid[3:, :2])
>>> ax1.plot(degree_sequence, "b-", marker="o")
>>> ax1.set_title("Degree Rank Plot")
>>> ax1.set_ylabel("Degree")
>>> ax1.set_xlabel("Rank")

>>> ax2 = fig.add_subplot(axgrid[3:, 2:])
>>> ax2.bar(*np.unique(degree_sequence, return_counts=True))
>>> ax2.set_title("Degree histogram")
>>> ax2.set_xlabel("Degree")
>>> ax2.set_ylabel("# of Nodes")

>>> fig.tight_layout()
>>> plt.show()
```

Output

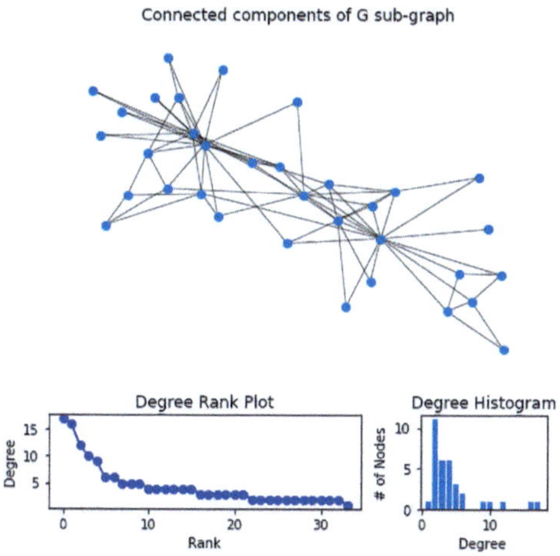

# Betweenness centrality
>>> **bet_cen = nx.betweenness_centrality(g)**
>>> **print(bet_cen)**

# Closeness centrality
>>> **clo_cen = nx.closeness_centrality(g)**
>>> **print(close_cen)**

# Eigenvector centrality
>>> **eig_cen = nx.eigenvector_centrality(g)**

# Compute the clustering coefficient of each node
>>> **print(nx.clustering(g))**

# This returns clustering value of specified node (in our case node 1)
>>> **print(nx.clustering(g, 1))**

# Compute the average clustering coefficient for the graph g
>>> **print(nx.average_clustering(g))**

Compute graph transitivity, the fraction of all possible triangles present in g **print(nx.transitivity(g))**

$$Transitivity = 3\frac{\#\text{Triangles}}{\#\text{Triads}}$$

# Returns the PageRank score for each node
>>> **PR = nx.pagerank(g, alpha = 0.6)**
>>> **print(PR)**

# Compute degree assortativity of graph. Assortativity measures the similarity of connections in the graph with respect to the node degree
>>> **assortativity = nx.degree_assortativity_coefficient(g)**
>>> **print(assortativity)**

# Returns the average degree of the neighborhood of each node.
>>> **nx.average_neighbor_degree(g, weight="weight")**

# Returns the constraint on all nodes in the graph g
>>> **cons = nx.constraint(g)**

# Returns the local constraint on node u in the graph g with regard to node v.
>>> **l_cons = nx.local_constraint(g, 1, 2)**

```
# Provides the effective size of all nodes in the graph g
>>> eff_size = nx.effective_size(g)

# Compute the Jaccard Coefficient of all node pairs and create an iterator of 3-tuples in the form (u, v, p), where (u, v) is a pair of nodes and p is their Jaccard Coefficient
>>> Jc = nx.jaccard_coefficient(g)
>>> for u, v, p in Jc:
>>>     print(f"({u}, {v}) -> {p:.3f}")

# Returns the Rich-Club coefficient of the graph
>>> rc = nx.rich_club_coefficient(g, normalized=False, seed=1234)
>>> print(rc)
```

3.10 Case Study Using Stata: Innovation Networks through Patent Data

Network data can be observed at various times. This view allows one to track the development of network edges over time. In this case study, we consider a sample of patent applications submitted to the European Patent Office. The European Research Infrastructure for Science, Technology, and Innovation Policy Studies, or RISIS, is the main source of our data. We examine the relationship between regions in 2012 and 2016 to see if there is a tendency to be open to less connected members throughout these 2 years, hence supporting inclusivity and integration within areas (Makkonen & Mitze, 2016; Balland et al., 2019). Furthermore, we look at the link between the innovation network and a region's major economic indicator, GDP per capita, to see whether there are any homophily effects. In this regard, the European Union (EU) has encouraged the establishment and growth of collaboration networks to integrate research resources and create a "network of excellence" in order to promote the European Research Area (ERA), a "common market" of EU research. The goal of the ERA is to increase the competitiveness of EU research institutions and regions (European Commission, 2020). In particular, the following research questions are addressed: (1) is there more regional integration in patent cooperation between 2012 and 2016? (2) Is there a GDP homophily effect on the likelihood of co-patent connections in the two studied years?

In both 2012 and 2016, at least one applicant with a European address was included in the data we used. Regarding the accuracy of the data, it is important to note that it is theoretically feasible that some applicants may have submitted their applications directly to other nations' patent offices, leaving them out of our database. However, these situations are extremely uncommon because the idea would not be protected by a patent in the applicant's country. Therefore, we have a sample of co-patent linkages between two European NUTS2 areas (248) for the years 2012 and 2016. When at least two inventors from distinct regions (regardless of the country) are listed in the same patent document, there is a link between the two regions. By using network principles, we created an undirected network that illustrates the inter-regional patent-filing relationships from the inventor regions, where vertices represent the regions and links show the interaction between the regions.

The data structure is in edge list format and for a total of 249 regions from 22 European countries (Austria, Belgium, Bulgaria, Czech Republic, Germany, Denmark, Greece, Spain, Finland, France, Hungary, Croatia, Ireland, Italy, Netherlands, Poland, Portugal, Romania, Sweden, Slovenia, Slovakia, and United Kingdom), V_i indicates the *ith* region, and an adjacency matrix $A = [a_{ij}]$ indicates the patent collaboration between regions, where $a_{ij} = n$ indicates the patent collaboration between regions, and $a_{ij} = 0$ otherwise. Self-loops (i.e., links connecting iii with itself) are not considered. Based on the co-patenting network, we have constructed the networks among all regions for the years 2012 and 2016. A patent with only two inventors relates to a single relational event (= one joint patent), while a patent with five inventors, for example, results in five pairwise relational ties (= five inventor pairs with a joint patent). Increased link intensity considers the fact that various inventor patents will result in multiple relational ties.

Together with information on links between regions (provided by a source and a target region along with the weight of the link), the dataset includes the following variables:

- **Country**: captures the presence of a country effect, indicating whether two regions belonging to the same country collaborate significantly more than two regions belonging to different countries.
- **Capital**: a binary variable indicating whether or not the region contains the country's capital (1 = capital; 0 = no capital).
- **GDP per capita**: a proxy for economic development of regions.
- **GERD**: indicates the gross domestic expenditure on Research and Development (R&D).
- **Geo distance**: measures the geographical proximity, computed as the distance in kilometers between the centroids of each region.

In this case study, we show how to generate network data in Stata. This case study also demonstrates how to use the *nwcommands* package (Grund & Hedström, 2015) to visualize and describe the graphs. The first line in the next code block generates several graphs based on the levels of the variable "year". The graphs we want to analyze are saved after running *foreach* through all possible values of the variable "year" and using the *nwset* command to create the graphs.

Input

```
. clear all
. nwset, nwclear
. cd "~\MyDirectory"
. use co-patents, clear
. levelsof year, local(levels) //levelsof is used to produce a list of the distinct values in a variable
. foreach y of local levels {
preserve
keep if year ==`y'
nwset source_L target_L weight , undirected edgelist labsfromvar(nuts2) name(net_`y')
save `y', replace
restore
}
```

In the following code block, we set the working directory and data path for the network file that was just generated and perform the basic centrality measures. We investigate co-patent networks for the years 2012 and 2016, as mentioned above.

Input

```
. clear all
. local net 2012 2016
. foreach data of local net{
use `data'
rename _nodelab nuts
nwset net*, undirected name(`data')
nwcurrent
nwsummarize, detail
}
```

Output

Network name: net_2012 Network id: 1 Directed: false Nodes: 249 Edges: 4,659 Minimum value: 0 Maximum value: 14 Density: 0.151 Reciprocity: 0.151 Transitivity: 0.455 Betweenness centralization: 0.054 Degree centralization: 0.470	Network name: net_2016 Network id: 1 Directed: false Nodes: 249 Edges: 5,461 Minimum value: 0 Maximum value: 14 Density: 0.177 Reciprocity: 0.177 Transitivity: 0.388 Betweenness centralization: 0.040 Degree centralization: 0.464

From the previous output, it is possible to see that, despite the same number of regions for the two years, the number of ties increases between 2012 and 2016. The density, namely the number of observed ties over all possible number of ties in the observed network, is slightly higher in 2016 than it was in 2012. Transitivity, betweenness, and degree centralization are higher in 2012. To summarize, the network as a whole has grown in terms of collaborations, but the network has become more decentralized and open, with multiple regions serving as centralized hubs for co-patent collaborations. Moreover, the absence of a closed group of innovators among regions, as indicated by the lowest transitivity value in 2016, suggests that innovation concentration can still be compatible with a sustainable and cohesive territorial development strategy in the EU.

The distribution of the variables degree, betweenness, closeness, farness, nearness, and clustering coefficient is implemented in the following code block. In addition, we use the *merge* command to combine the newly generated datasets *measures_net_2012* and *measures_net_2016* with the original dataset *co-patents*.

Input

```
. clear all
. local net 2012 2016
. foreach data of local net{
use `data'
rename _nodelab nuts
nwset net*, undirected name(`data')
nwcurrent
gen density=r(density)
gen transitivity= r(transitivity)
nwdegree, gen(degree) standardize valued
nwbetween, gen(betweenness) standardize
nwcloseness, generate(closeness farness nearness) unconnected(1)
nwclustering, gen(cluster_coefficient)
save `data', replace
}
. clear all
. local net 2012 2016
. foreach data of local net{
use `data'
gen year=`data'
merge 1:m nuts year using co-patents.dta
drop if _merge!=3
drop _merge
save Net_`data', replace
}
```

Then, in the following box, we generate the degree map for the years 2012 and 2016 using the *shp2dta* command, which converts shape boundary files to Stata datasets.

Output

Input

```
. clear all
. local network Net_2012 Net_2016
. foreach data of local network{
        preserve
 use `data'
 drop if _nodeid==.
 keep _nodeid source_L source nuts year degree bet closeness clu density transitivity
 shp2dta using geom_eu34_nuts13, data(d_`data')  coor(c_`data') replace
          restore
 }
. clear all
. local network Net_2012  Net_2016
. foreach data of local network{
 use d_`data',clear
 rename id nuts
 merge m:m nuts using `data'.dta
 drop if _ID==.
 drop if year==.
 duplicates drop _ID , force
 format degree %9.2f
 spmap degree using c_`data'.dta, id(_ID) clmethod(q) clnumber(4) fcolor(Heat) legorder(lohi)
 osize(thin)  /// note("Source: Authors' elaboration") legend(ring(1) position(3) size(*0.9)) ///
 legtitle("Quantile classification") title("Degree centrality")
 graph save `data'_map.gph, replace
 }
. graph combine Net_2012_map.gph Net_2016_map.gph, ycommon xcommon commonscheme ///
  imargin(zero) ysize(2.5) graphregion(color(white) lwidth(large))
```

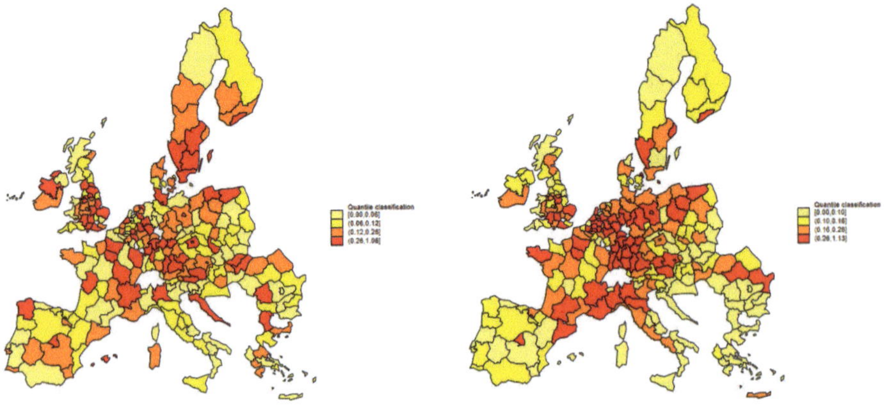

Fig. 3.12 Degree centrality maps for 2012 (left) and 2016 (right)

The maps reported in Fig. 3.12 show regional Degree centrality for 2012 (left) and for 2016 (right) for co-patent collaboration. By comparing the static snapshots in these 2 years two facts emerge. Firstly, there is some significant degree of overlapping among the two considered years. In 2016, various areas in the belt stretching from Northern Italy, to the core of the German manufacturing industry, together with Denmark, Austria, France, Poland, and the capital region of the UK, are consistently in the higher quartile of the Degree centrality. The second fact is the presence of several regions that lost importance between 2012 and 2016 (some regions of Norway and Ireland, for example).

To summarize, Fig. 3.2 shows that certain areas have lost centrality in patent collaborations, whereas others have gained centrality by becoming innovation hubs and attracting many peripheral regions to collaborate.

Figure 3.13 illustrates several triad orientations that can be created between nodes using unidirectional and bidirectional linkages. There are a total of 16 possible combinations, each assigned an identification code to facilitate understanding of the quantity and type of links represented in this basic graphical representation of a network.

Table 3.4 displays the triads for both networks 2012 and 2016. Given the undirected nature of our network data, certain configurations are not applicable. Therefore, we categorize the two networks based on the possible triads that can occur.

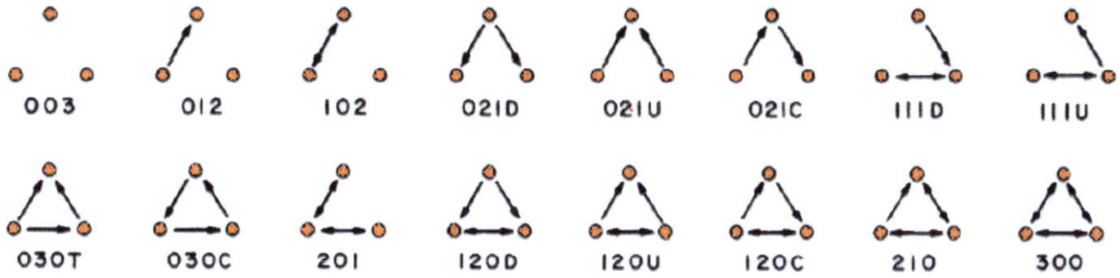

Fig. 3.13 Triad configurations

Table 3.4 Triad configurations for 2012 and 2016

Co-patent network 2012				Co-patent network 2016			
003	**012**	**021D**	**021 U**	**003**	**012**	**021D**	**021 U**
1,653,020	0	0	0	1,492,472	0	0	0
021C	**030 T**	**030C**	**102**	**021C**	**030 T**	**030C**	**102**
0	0	0	674,197	0	0	0	794,864
120D	**120 U**	**120C**	**111D**	**120D**	**120 U**	**120C**	**111D**
0	0	0	0	0	0	0	0
111 U	**201**	**210**	**300**	**111 U**	**201**	**210**	**300**
0	168,145	0	46,762	0	210,361	0	44,427

Considering that in the two networks there is no direction of edges, we conducted our analysis looking for variation in the frequency of four specific triad configurations, namely the configurations with the codes: 003, 102, 201, and 300. In 2016, there are fewer empty triads (code 003) in absolute terms, indicating increased linkages between European regions. This is further highlighted by the greater frequency of "102" and "201" codes in the 2016 network. Finally, because the "300" code is less common in 2016, small and densely interconnected cliques play a less crucial role in the construction of these networks. As a result, the 2016 network is more open to collaboration, with fewer so-called closed clubs.

The next box describes how to use Stata to create a scatter plot in Python using the three centrality measures, degree, betweenness, and closeness centrality. Regions are identified by distinct colors based on GDP per capita levels (low, medium, and high).

Input

```
. clear all
. use Net_2012
. python:
import os
os.environ['QT_QPA_PLATFORM_PLUGIN_PATH']= "C:/Users/anaconda3/Library/plugins"
from sfi import Data
end
. python:
from __future__ import division
import numpy
import numpy as np
from mpl_toolkits.mplot3d import Axes3D
import matplotlib.pyplot as plt
numpy.seterr(divide='ignore', invalid='ignore')
import pandas as pd
from sklearn import datasets
import seaborn as sns, numpy as np
import matplotlib.pyplot as plt
data = pd.read_stata('Net_2012.dta')
print(data)
sns.pairplot(data,
kind="scatter",palette="tab10",markers=["s","o","D"],size=2.5,aspect=1.8,diag_kind="kde",vars=["
Degree","Betwenness","Closennes"],hue="Gdp_per_capita",diag_kws=dict(shade=True))
plt.show()
end
```

For the year 2012, the following scatter plot (Fig. 3.14) displays the uniform distribution along the diagonal as well as the relationships between the three measures off the diagonal.

Output

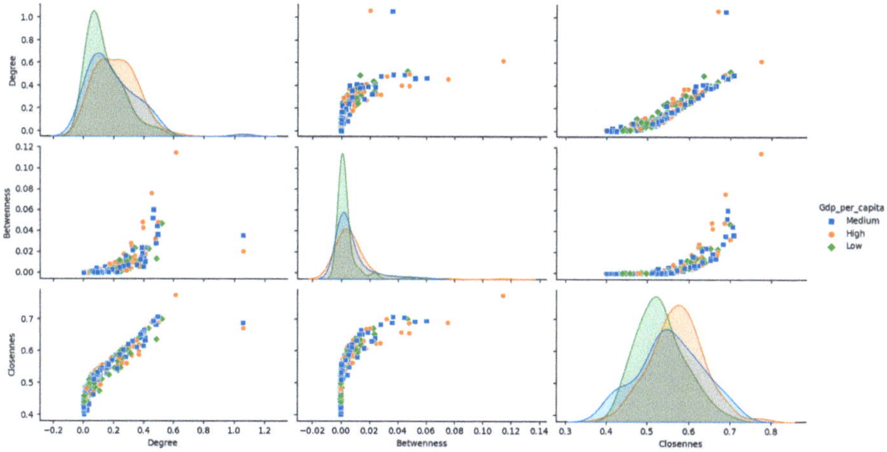

Fig. 3.14 Univariate distributions of three centrality measures (on the diagonal) along with all their pairwise relationship for 2012

We can see from the degree centrality, which is the first uniform distribution, that a large percentage of the regions with an average GDP make up the tail on the right of the distribution. Therefore, regions with an average GDP have the highest degree in 2012. In terms of betweenness and closeness, the regions with the highest GDP are located in the far right tail of the uniform distribution.

Furthermore, we observe substantial point overlap when viewing the pairwise relationship off the diagonal. As a result, there is no detectable clustering of regions based on their GDP.

In Fig. 3.15, we show the scatter plot for the year 2016. As with the previous figure, we display the distribution on the diagonal as well as the relationship between the three measures off the diagonal.

Output

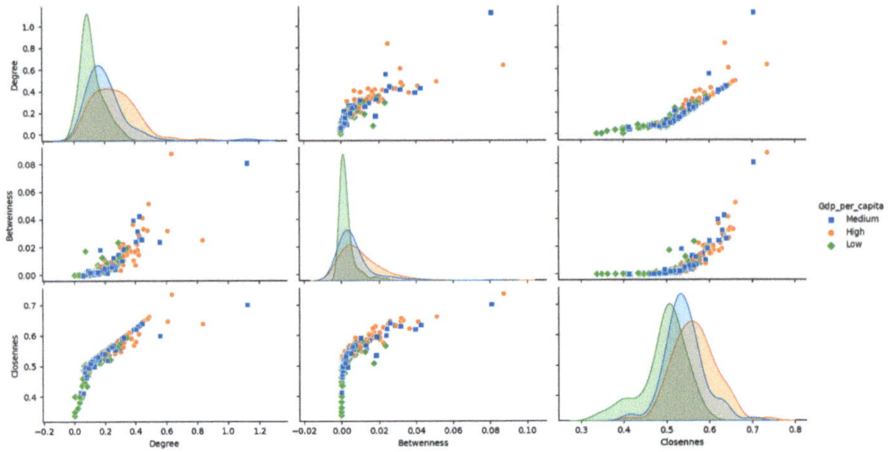

Fig. 3.15 Univariate distribution of the three centrality measures (on the diagonal), along with all their pairwise relationship for 2016

Figure 3.15 shows the existence of a very strong positive and structural correlation among the degree, betweenness, and closeness centrality. A high positive structural dependence between these centrality measures is evident from the pairs of analyzed centrality measures' strong positive correlation. However, there is a significant positive and nonlinear relationship between betweenness centrality and closeness centrality. This nonlinear relationship reflects the different levels of the GDP indicator across regions. Especially in 2016, the three groups of regions identified by the GDP variable with respect to the centrality measures are clearly visible.

The two co-patent networks are graphically shown in Fig. 3.16, where each node is a region, and the lines represent the existence of one link. We do not consider isolated nodes and weighted links for simplicity's reasons.

Output

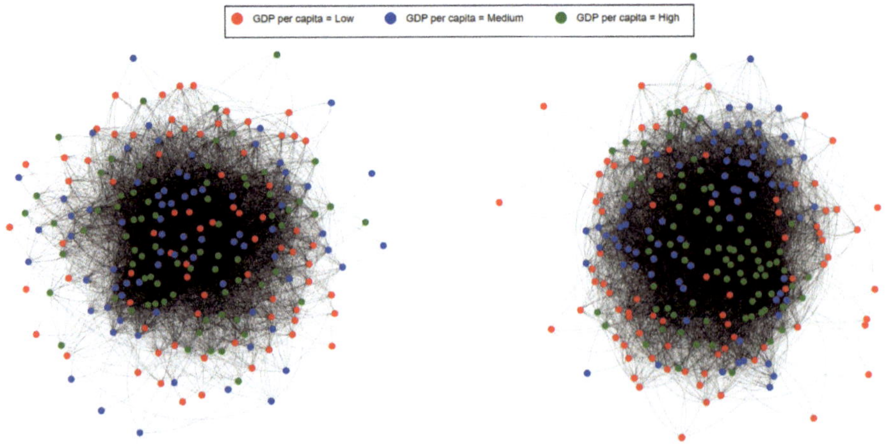

Fig. 3.16 Collaborative innovation graph for 2012 (left) and 2016 (right)

One can see a decrease in the number of peripheral nodes over time, which reflects a dynamic of growing integration among the European regions throughout the time period under consideration. It is clear that the nodes at the graph's extreme boundaries in 2016 compared to 2012 are all regions with low GDP (red circle). Always in 2016, regions with high GDP are mostly located in the center of the graph.

Looking at the latter three figures together (Figs. 3.14, 3.15, and 3.16), one can observe that in 2016, there is a definite distinction of regions in terms of GDP and network centrality. We stress the significance of regional GDP in the network's position in 2016 much more than in 2012.

Next tables (Tables 3.5 and 3.6) illustrate the two-way table for the years 2012 and 2016. The two tables display the number of connections between regions that have a specific level of GDP per capita (low, medium, and high). This is useful for detecting homophily in the two co-patent networks. We are particularly interested in determining whether there is a tendency for edges to exist between similar nodes in terms of GDP per capita.

Table 3.5 displays that out of 4659 links in the co-patent network for the year 2012, 268 links are between two regions who have both a low GDP per capita, 632 are between two regions who have both a medium level of GDP per capita and 758 links are between two regions who have an high GDP per capita. Outside the diagonal are the connections among regions with varying levels of GDP per capita. The E-I index has a substantial value of 0.29, indicating low heterophily between regions for the indicator GDP per capita.

Table 3.6 shows that among the 5461 links in the co-patent network for 2016, 216 are between two regions with low GDP per capita, 873 are between two regions with a medium GDP per capita, and 1384 are between two regions with a high GDP per capita. Outside the diagonal are the connections among regions with varying

Table 3.5 The frequency of edges between regions at the GDP per capita levels for 2012

From GDP per capita 2012	To GDP per capita 2012				
		Low	Medium	High	Total
	Low	268	590	315	1173
	Medium	167	632	171	970
	High	612	1146	758	2516
	Total	1047	2368	1244	4659
	E-I Index: 0.29 p-value: 0				

Table 3.6 The frequency of edges between regions at the GDP per capita levels for 2016

From GDP per capita 2016	To GDP per capita 2016				
		Low	Medium	High	Total
	Low	216	411	283	910
	Medium	225	873	253	1351
	High	640	1176	1384	3200
	Total	1081	2460	1920	5461
	E-I Index: 0.09 p-value: 0				

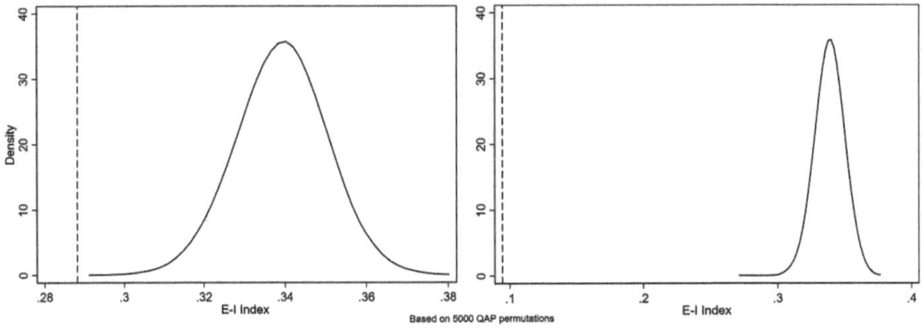

Fig. 3.17 Observed E-I index vs simulated E-I index in 2012 (left) and 2016 (right)

levels of GDP per capita. The E-I index has a significant value of 0.09, indicating no homophily between areas for the variable GDP per capita.

The increase in the overall number of links, coupled with the decline in the E-I index between 2012 and 2016, supports our earlier inference that there is more interaction and openness among regions.

The results of the permutation test, which determine whether the E-I index value significantly differs from what would be predicted by random mixing, are shown in Fig. 3.17. The number of internal and external links at various levels in the GDP per capita variable is estimated as a sampling distribution. By selecting from a population in which links were randomly distributed, we may determine how frequently the observed outcome would occur using the sampling distribution. We have performed 5000 trials for each network. The dotted line is the E-I index calculated on the observed network, the distribution is a representation of permutated networks.

We can see that the random test returns a higher value than the observed value for both years. These can be used as p-values and are expressed as probabilities. The significant difference between the dotted lines and the distributions suggests that the E-I values for 2012 (0.29) and 2016 (0.09) are significant.

3.11 Conclusion

Our analysis reveals the increasing integration of the European innovation collaboration network. This is evidenced by the fact that while the number of links between regions increased between 2012 and 2016, the number of triangles decreased. This is solid evidence that the regions have become more integrated over the course of the 2 years under study. The findings suggest that regions play a dual role: higher GDP regions are increasing collaboration with regions of medium and lower GDP, while also strengthening their central role within the network, thereby enhancing the presence of a network of great innovators.

These results support the on-going discussion on economic convergence among European regions. It is possible that the growing importance of richer regions and their innovation network is causing economic divergence to worsen, but the integration of other regions with a lower GDP gives us reason to be hopeful about the possibility of a convergence process. Our findings agree with two major trends in regional STI patterns: inequality in the distribution of technological capabilities and the persistence of innovation in the more developed regions, as well as some technological convergence between the less developed peripheral regions of Europe and the more developed central regions of the EU (Filippetti et al., 2020).

We are aware that innovation frequently occurs in fields and ways that are only partially covered by patents, and that patents only capture a portion of innovation. This case study has contributed to understanding the phenomenon of co-patenting among regions, such as their patterns of growth over time and their internal dynamics of disparities and inequalities. From a policy perspective, this study's findings have shown that the integrated network model is a desirable aspect of economic development because it suggests a knowledge-sharing and innovation mechanism.

References

Balland, P. A., Boschma, R., & Ravet, J. (2019). Network dynamics in collaborative research in the EU, 2003–2017. *European Planning Studies, 27*(9), 1811–1837.

Barabási, A. L. (2016). *Network science*. Cambridge University Press.

Borgatti, S. P. (1997). Structural holes: Unpacking Burt's redundancy measures. *Connect, 20*(1), 35–38.

Brin, S., & Page, L. (1998). The anatomy of a large-scale hypertextual web search engine. *Computer Networks and ISDN Systems, 30*(1–7), 107–117.

Burt, R. S. (1991). Measuring age as a structural concept. *Social Networks, 13*, 1–34.

Burt, R. S. (1992). *Structural holes*. Harvard University Press.

Burt, R. S. (2000). The network structure of social capital. *Research in Organizational Behavior, 22*, 345–423.

Colizza, V., Flammini, A., Serrano, M. A., & Vespignani, A. (2006). Detecting rich-club ordering in complex networks. *Nature Physics, 2*(2), 110–115.

De Andrade, R. L., & Rêgo, L. C. (2018). The use of nodes attributes in social network analysis with an application to an international trade network. *Physica A: Statistical Mechanics and its Applications, 491*, 249–270.

De Andrade, R. L., & Rêgo, L. C. (2022). A proposal for the EI index for fuzzy groups. *Soft Computing*, 1–13.

De Domenico, M., & Arenas, A. (2017). Modeling structure and resilience of the dark network. *Physical Review E, 95*(2), 022313.

European Commission. (2020). *A new era for research and innovation*. COM(2020) 628 final.

Everett, M. G., & Borgatti, S. P. (2012). Categorical attribute based centrality: E-I and G-F centrality. *Social Networks, 34*, 562–569.

Filippetti, A., Gkotsis, P., Vezzani, A., & Zinilli, A. (2020). Are innovative regions more resilient? Evidence from Europe in 2008–2016. *Economics and Politics, 37*, 807–832. https://doi.org/10.1007/s40888-020-00195-4

Granovetter, M. S. (1973). The strength of weak ties. *American Journal of Sociology, 78*(6), 1360–1380.

Grund, T., & Hedström, P. (2015, September). Social network analysis using Stata. In *Proceedings of the United Kingdom Stata Users' Group Meeting*.

Jackson, M. (2010). *Social and economic networks*. Princeton University Press.

Jiang, Z. Q., & Zhou, W. X. (2008). Statistical significance of the rich-club phenomenon in complex networks. *New Journal of Physics, 10*(4), 043002.

Katz, L. (1953). A new status index derived from sociometric analysis. *Psychometrika, 18*(1), 39–43.

Krackhardt, D., & Stern, R. N. (1988). Informal networks and organizational crises: An experimental simulation. *Social Psychology Quarterly, 51*, 123–140.

Makkonen, T., & Mitze, T. (2016). Scientific collaboration between 'old' and 'new' member states: Did joining the European Union make a difference? *Scientometrics, 106*(3), 1–23.

Newman, M. (2018). *Networks*. Oxford University Press.

Resce, G., Zinilli, A., & Cerulli, G. (2022). Machine learning prediction of academic collaboration networks. *Scientific Reports, 12*, 21993. (2022). https://doi.org/10.1038/s41598-022-26531-1

Watts, D. J., & Strogatz, S. H. (1998). Collective dynamics of small-world networks. *Nature, 393*, 440–442.

Zhou, S., & Mondragón, R. J. (2004). The rich-club phenomenon in the Internet topology. *IEEE Communications Letters, 8*(3), 180–182.

Zinilli, A., & Cerulli, G. (2023). Link prediction and feature relevance in knowledge networks: A machine learning approach. *PLoS One, 18*(11), e0290018.

Chapter 4
Theoretical Models of Networks

4.1 Introduction

Complex systems are characterized by numerous nonlinearly interacting components. These components give rise to emergent characteristics that challenge linear assumptions. In the domain of social and economic sciences, a fundamental challenge lies in understanding how the interactions and characteristics of individual entities within a social or economic system contribute to the overall behavior and dynamics of the system as a whole. It involves analyzing how the attributes, such as beliefs, preferences, actions, or economic factors, of the individual components influence the emergent properties and patterns observed at the collective level. As previously mentioned in the initial chapter, these complex interconnections have led to the recognition of the analysis of complex networks as a vital tool in the investigation of social and economic phenomena. Once a social or economic phenomenon is translated into a network representation, the tools provided by complex network theory facilitate a systematic examination of its structural properties. When we analyze network topology, we aim to understand how social and economic systems are structured, how they work, and what principles govern their organization. This exploration helps us uncover the underlying design and organizational principles that were not previously known. By studying the connections and relationships between components within these networks, we gain insights into how these systems work and what drives their behavior. Recently, network scientists from the natural sciences have noticed something interesting. They have discovered that complex networks, found in various areas like biology, sociology, and economics, often share similar patterns (Albert & Barabási, 2002; Newman, 2003). These patterns in complex networks show us some important characteristics of the systems they represent. For example, they can tell us how well the systems can recover from

A. Zinilli, *Elements of Network Science*, Statistics and Computing,
https://doi.org/10.1007/978-3-031-84712-7_4

difficulties, if there are alternative options available, and how different parts are connected to each other. Understanding these qualities helps us see how resilient and adaptable these systems are in the face of challenges. We can learn about how social and economic systems are structured by using specific measures to study their overall features. This gives us a better understanding of how different parts fit together and how the systems work as a whole. Although this chapter does not aim to cover everything, its main goal is to give a concise overview and explanation of some important characteristics and traits related to network topology. It seeks to provide a clear understanding of the fundamental aspects of how networks are structured and organized.

In this section, we will discuss some of the most significant topological properties: the small-world effect and the power-law distribution (also known as scale-free), which have been observed in recent social studies. To achieve this, however, it is crucial to first introduce the model that has long served as the reference for representing networks: the random network. Subsequently, more recently developed models that are capable of reproducing certain observed characteristics in real networks will be presented. By comparing these models, we can more easily deduce some of the properties of the more advanced ones.

4.2 Erdös–Rényi Random Networks

The initial key distinction in network topology lies between structured and random networks. Structured networks are characterized by a certain level of regularity in their connectivity patterns, with nodes having a specific arrangement of connections that often follow a predefined set of rules or patterns. This organization leads to local clustering and predictable connectivity, with each node typically having a similar number of connections resulting in strong interconnection among nodes. In contrast, random networks evolve through stochastic processes, with structures that do not exhibit the same level of organization. Structured networks tend to have high clustering coefficients and, compared to random networks, longer average path lengths between arbitrarily chosen nodes. However, they are less adept at capturing the complexity and heterogeneity of phenomena like social interactions, in which diverse individuals or groups interact in a variety of ways. On the other hand, in a random network, the edges between vertices are typically established without any specific rules or patterns. The connections are formed randomly, without considering any inherent organization or coordination. The classical mathematical model was formulated by Erdös and Rényi in 1960. In their model, a network is constructed by randomly connecting nodes from a given set, with each pair of nodes having a fixed probability, denoted as p, of being linked. It can be mathematically proven that in these random networks, the degree distribution follows a Poisson distribution. The characteristics of a Poisson distribution imply that many nodes in a random network have a similar number of edges, giving rise to a certain degree of homogeneity. Despite the random assignment of edges, a typical random graph tends to have a narrow degree distribution that closely approximates the average value. Additionally, the path lengths between nodes in random networks are generally short, but due to the independent nature of edges, these networks lack localized topological structures. However, it is important to acknowledge that the Erdös and Rényi model, despite its foundational role, has limitations in accurately capturing the complexity of real-world networks. While the random model provides a useful baseline for understanding certain network characteristics, it fails to fully capture the intricate features and organizational principles observed in diverse real systems. Real networks, ranging from social networks to biological networks and technological networks, often exhibit properties that deviate from those predicted by the random model.

Suppose we have N individuals, and on average, each person has l connections. In the Erdős–Rényi model, the probability p of an edge existing between two vertices can be calculated as:

$$p^l (1-p)^{N(N-1)p/2-l} \tag{4.1}$$

The topology of the generated network varies across different realizations. Each of the $\frac{N(N-1)}{2}$ pairs of nodes is connected with a probability p, resulting in a mean number of edges $l = \frac{N(N-1)p}{2}$. The mean degree k of a node in the network is equal to $(N - 1)p$.

Now, we will investigate the degree distribution within the Erdős–Rényi model (Erdős & Rényi, 1960). Each node in the network connects to the other N-1 nodes with a probability p. The probability distribution $P(k)$, which represents the likelihood of a node having k neighbors, can be defined as follows:

$$P(k) = \frac{(N-1)!}{k!(N-1-k)!} p^k (1-p)^{N-1-k} \tag{4.2}$$

When we examine sparse networks in the limit of large size N approaching infinity ($N \to \infty$) while keeping the mean degree $k = (N - 1)p$ constant, the probability distribution $P(k)$ approaches a Poisson distribution in the limit of large N. Approximating the distribution $P(k)$ as a Poisson distribution in the limit of large N means that in an ordinary random graph, where there are a large number of nodes, the degree of the nodes in the graph tend to follow a Poisson distribution. A Poisson distribution is a probability distribution that describes the number of events occurring in a fixed interval of time, given a known average rate of occurrence and the assumption that the events happen independently and at a constant average rate. In the context of a random graph, the Poisson distribution characterizes the likelihood of observing a certain degree of connectivity *(k)* for a node:

$$\lim_{N \to \infty} P(k) \simeq \frac{e^{-k} (k)^{k}}{k!} \tag{4.3}$$

The behavior of a random graph is significantly influenced by the parameter p. When we consider a network generated using specific values of N (the size of the network) and k (the mean degree of the nodes), it becomes important to investigate whether this network is likely to have a giant connected component. As defined in the third chapter of this book, a giant connected component refers to a part of the network where a significant portion of the nodes is interconnected, forming a large connected subgraph. If a network has a giant connected component, its size is proportional to N, meaning that as the network size increases, the size of the giant connected component also increases.

The clustering coefficient C tends to approach zero as the network size N increases. This occurs because the probability of a connection between two nodes is p, irrespective of whether they share a common neighbor. Hence, the clustering coefficient C can be approximated as:

$$C = p \simeq \frac{k}{N} \tag{4.4}$$

As the network size N grows, the clustering coefficient C tends to decrease toward zero. The Eq. 4.4 highlights a significant characteristic of the Erdős–Rényi model: in the case of a large network size, the model exhibits a local tree-like structure. This suggests that within a local neighborhood, there are few or no cycles or closed loops. Nodes within this local region tend to have fewer direct connections to each other, with connections more likely forming a tree-like branching hierarchical structure. However, it is crucial to note that this local tree-like structure does not imply that the entire network is a tree. When considering the network as a whole, the overall structure can still exhibit loops and cycles. The local tree-like structure mainly refers to the connectivity pattern observed within a node's limited neighborhood. This implies that finite cyclic paths become negligible when N is greater than or equal to 1. Consequently, k^l represents the mean number of neighbors at a distance of l away from a randomly chosen node. If this pattern continues until all neighbors encompass the entire network, k^l approaches N, indicating that each node is connected to most other nodes in the network.

Furthermore, Equation 4.5 provides insights into the typical distance within the Erdős–Rényi model. It reveals that the typical distance is

$$l \simeq \frac{\log N}{\log(k)} \tag{4.5}$$

This Eq. (4.5) suggests that the Erdős–Rényi model exhibits small-world characteristics, meaning that even in large networks, the average distance between nodes remains relatively small compared to the network size. The Erdős–Rényi model shows small-world characteristics in terms of average distance, but it might not fully represent all the properties found in real small-world networks. The Erdős–Rényi model often does not account for additional characteristics that are frequently seen in real networks, such as the presence of hubs and node clustering. However, understanding the small-world concept helps us grasp the underlying structure and dynamics of various complex systems. In the Sect. 4.4, we will delve into the concept of the small world and its implications in the social and economic sciences.

4.2.1 Stata Code and Output

```
// Clear any existing network data
. nwclear

// Generate a random graph with 50 nodes and an edge probability of 0.1
. nwrandom 50, prob(0.1)

// Generate a random graph with 15 nodes and a density of 0.5
. nwrandom 15, density(0.5)

// Generate 5 random graphs with 20 nodes and an edge probability of 0.3
. nwrandom 20, prob(0.3) ntimes(5)

// Generate a random undirected graph with 10 nodes and an edge probability of 0.2
. nwrandom 10, prob(0.2) undirected

// Summarize the network data
. nwsummarize _all

// Plot the random graph
. nwplot random_2
```

Output

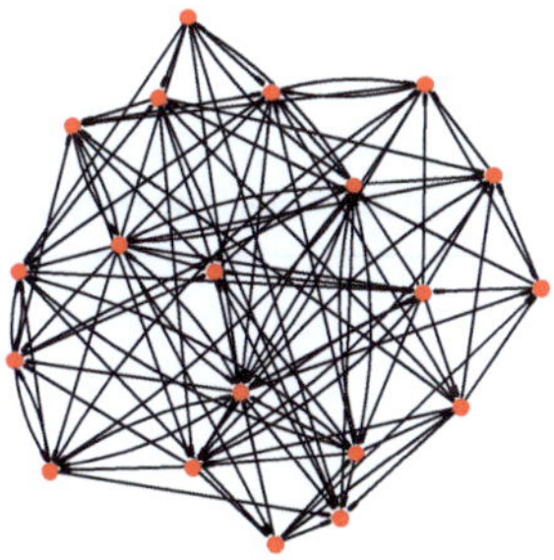

4.2.2 R Code and Output

```
# Load the igraph package
> library(igraph)

# Set the random seed for reproducibility
> set.seed(123)

# Generate a random graph with 18 nodes and an edge probability of 0.15
> random_graph <- erdos.renyi.game(18, p = 0.15)

# Define a color palette for the nodes
> vertex_colors <- c("#FF7F00", "#377EB8", "#4DAF4A", "#984EA3", "#E41A1C")

# Generate alphabetic labels for the nodes
> labels <- LETTERS[1:18]

# Plot the random graph
> plot(random_graph,
    layout = layout_with_kk, # layout function
    vertex.size = 10,
    vertex.color = vertex_colors,
    vertex.label = labels,
    vertex.label.color = "black",
    vertex.label.dist = 2, # increased label distance
    vertex.label.cex = 0.8,
    vertex.frame.color = "white",
    edge.color = "black",
    edge.width = 1,
    edge.curved = 0.2,
    edge.arrow.size = 0.5,
    main = "Random Graph",
    sub = "18 Vertices, Edge Probability 0.15",
    xlim = c(-1.5, 1.5),
    ylim = c(-1.5, 1.5))
```

Output

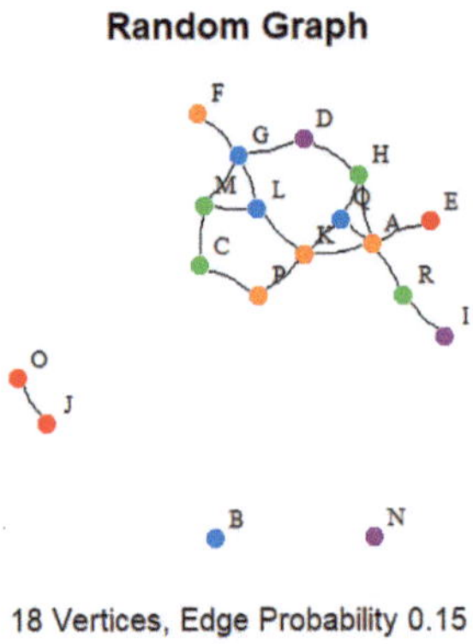

4.2.3 Python Code and Output

```
# Importing the required packages
>>> import networkx as nx
>>> import matplotlib.pyplot as plt
>>> import random

# Set the random seed for reproducibility
>>> seed = 123
>>> random.seed(seed)

# Generate a random graph with 50 vertices and an edge probability of 0.1
>>> random_graph = nx.erdos_renyi_graph(50, 0.1, seed=seed)

# Generate alphabetic labels for each node
>>> labels = [chr(65 + i) for i in range(50)]

# Generate random colors for each vertex
>>> colors = [random.choice(['#FF7F00', '#377EB8', '#4DAF4A', '#984EA3', '#E41A1C']) for _ in range(50)]

# Create a figure and axis for the plot
>>> fig, ax = plt.subplots()

# Plot the random graph
>>> pos = nx.spring_layout(random_graph, seed=seed)
>>> nx.draw_networkx(random_graph, pos=pos, ax=ax, with_labels=True,
         node_color=colors, node_size=300,
         font_color='black', font_size=10,
         edge_color='gray', width=1, alpha=0.8)

# Set axis limits
>>> ax.set_xlim(-1.5, 1.5)
>>> ax.set_ylim(-1.5, 1.5)

# Add a title to the plot
>>> ax.set_title("Random Graph", fontsize=16)

# Show the plot
>>> plt.show()
```

Output

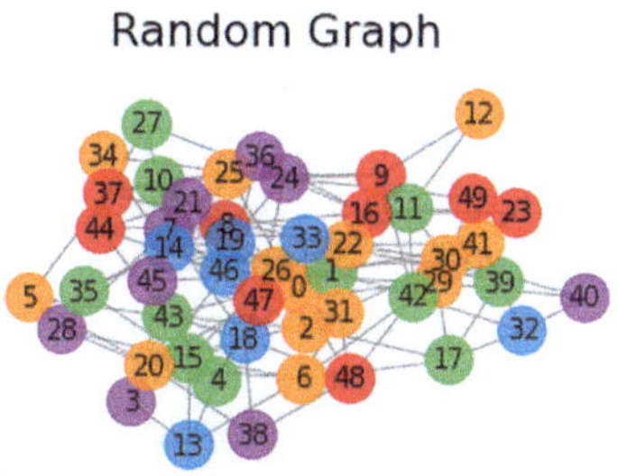

4.2.4 Community Structures in Random Networks

Real-world networks, including social and economic networks, often exhibit a modular organization characterized by the presence of communities (Girvan & Newman, 2002; Fortunato & Newman, 2022). These communities are composed of groups of vertices that adhere to a simple rule: vertices within the same community have a relatively high number of interconnecting ties, while interconnections between different communities are relatively sparse. In order to generate networks with community structures, several methods have been proposed (Girvan & Newman, 2002; Newman, 2004; Rosvall & Bergstrom, 2008; Fortunato, 2010; Battiston et al., 2020). However, we often lack sufficient information about the quantity or nature of these community clusters. For example, the graph might display a layered or hierarchical structure, where smaller clusters are embedded within bigger ones. As a result, traditional graph partitioning techniques may not be the most effective in such situations.

The initial step in any hierarchical clustering method is to establish the degree of similarity between nodes. After selecting a suitable measurement method, we compute the similarity across all node pairs, regardless of their connection status. This leads to the formation of a "similarity matrix", an important tool used in hierarchical clustering methods. In a similarity matrix, each entry would have a row and a column, and the intersection of a row and column represents the similarity between two nodes. The higher the value at the intersection, the more similar those two nodes are. These values are computed using various similarity metrics, the selection of which can be contingent upon the specific application. This matrix then serves a central role in hierarchical clustering as it enables the identification of nodes that should be merged based on their degree of similarity. This classification can be executed through two primary techniques:

- Agglomerative algorithms, where each vertex is initially treated as an individual, distinct cluster. These stand-alone clusters are subsequently incrementally combined, depending upon their degree of similarity.
- Alternatively, divisive algorithms start by treating the entire network as a single component. The huge cluster is then gradually divided by removing the ties between the nodes that exhibit lower similarities.

A significant and commonly used algorithm in community detection is the method introduced by Girvan and Newman in 2002. This method employs an iterative process to incrementally form cohesive communities or clusters by uniting initially isolated vertices (Fortunato, 2010). At the beginning, the network lacks any connections, and each vertex is isolated. The process based on agglomerative methods starts by gradually forming communities or clusters from these isolated vertices. Then the specific model determines which vertices should be grouped together to form communities based on certain criteria or rules. As a result, the formation of communities involves creating ties between vertices based on certain probabilities.

If two vertices belong to the same community, a tie is formed with a probability p_{in}, whereas if they belong to distinct communities, a tie is established with a probability p_{out}. By adjusting the values of p_{in} and p_{out}, one can effectively control the density of intracommunity and intercommunity ties within the network. These quantities, represented as c_{in} and c_{out}, respectively, are modified in relation to the average network degree $\hat{k}$.

Networks with high p_{in} values and low p_{out} values are characterized by well-defined communities, exhibiting a dense concentration of ties within each community and limited interconnections between different communities. Conversely, networks characterized by low p_{in} values and high p_{out} values foster highly interconnected and mixed communities, where the boundaries between communities become less distinct, with a higher level of interconnections between different communities. These properties allow us to define the fraction of intracommunity ties $c_{in}/\hat{k}$ and the fraction of intercommunity ties $c_{out}/\hat{k}$. The $c_{out}/\hat{k}$ ratio serves as a measure of the level of mixture among different communities. As $c_{out}/\hat{k}$ increases, the communities become progressively more entangled, making their identification more challenging. Empirically, it is crucial to ensure $p_{out} \leq p_{in}$ to guarantee the presence of distinct communities within the network (Girvan & Newman, 2002; Fortunato, 2010).

Another important method is the Louvain community detection algorithm, which is a straightforward and efficient technique used to extract the community structure within a network. The Louvain community detection algorithm provides a robust approach to understanding the community structure present in complex networks. Optimizing the modularity measure, it efficiently identifies groups of vertices that are densely connected within themselves while having fewer connections to vertices outside their community. This ability to uncover meaningful communities is vital in various fields, such as in social sciences (Traag et al., 2019). This heuristic method is primarily based on the optimization of modularity, a measure that quantifies the strength of the division of a network into communities. The process of the algorithm can be broken down into two main stages. In the first stage, each vertex of the network is initially designated to be its own community. Subsequently, the algorithm assesses each vertex and seeks to ascertain whether there is a positive modularity gain that can be realized by shifting the vertex to any of its neighboring communities. If the movement of the vertex does not result in a positive gain, the vertex remains within its original community. Therefore, when considering moving a vertex from its current community to a different one (a neighboring community), it first calculates whether this move would result in a positive gain, i.e., an increase in the overall modularity of the network. This strategy helps to maximize the overall modularity of the network and results in a more accurate division into communities.

The equation to calculate the modularity gain derived from moving an isolated vertex into a community is as follows:

$$\Delta Q = \frac{k_{i,in}}{2m} - \gamma \frac{\sum_{tot} k_i}{2m^2} \tag{4.6}$$

In Eq. 4.6, m stands for the size of the graph, $k_{i,\,in}$ represents the sum of the weights of the edges from the generic vertex i to vertices in a community (C), k_i is the sum of the weights of the edges incident to vertex i, Σ_{tot} is the sum of the weights of the edges incident to vertices in C, and γ is the resolution parameter (Blondel et al., 2008; Traag et al., 2019).

The first phase continues until no individual node movement can enhance modularity. After this stage, the second phase begins by constructing a new network where the vertices are the communities identified in the initial phase. The weights of the edges between the new vertices are assigned based on the sum of the weight of the edges between vertices in the two corresponding communities. Following this stage, the first phase can be used again to create larger communities that boost modularity.

These two phases are repeatedly executed until the modularity gain is negligible or less than a defined threshold. The repeated execution of the two phases ensures that the algorithm iteratively fine-tunes the partitioning of the network, striving to achieve an optimal division that maximizes modularity. This iterative nature provides flexibility in capturing complex community structures, even in large networks with a multitude of partitions.

In the next box, the generated communities exhibit clearly defined boundaries and are distinct from one another.

4.2.5 R Code and Output

```
# Importing the required libraries
> library(igraph)
> library(ggplot2)

# Set the number of nodes per community and the number of communities
> nodes_per_community <- 20
> num_communities <- 4

# Create an empty graph
> G <- make_empty_graph(n = nodes_per_community * num_communities)

# Add nodes to the graph
> V(G)$name <- 1:(nodes_per_community * num_communities)

# Add edges within communities
> for (c in 1:num_communities) {
  community_nodes <- ((c - 1) * nodes_per_community + 1):(c * nodes_per_community)
  G <- add_edges(G, combn(community_nodes, 2))
}

# Add a few intercommunity edges
> inter_edges <- 10
> for (i in 1:inter_edges) {
  source_community <- sample(1:num_communities, 1)
  target_community <- sample(setdiff(1:num_communities, source_community), 1)
  source_node <- sample(((source_community - 1) * nodes_per_community + 1):(source_community * nodes_per_community),
1)
  target_node <- sample(((target_community - 1) * nodes_per_community + 1):(target_community * nodes_per_community),
1)
  G <- add_edges(G, c(source_node, target_node))
}

# Generate a layout for the graph
> layout <- layout_with_fr(G)


# Assign pastel colors to each community
> colors <- c("#FFA500", "#FF0000", "#00FF00", "#0000FF")

# Set the size of the nodes and width of the edges
> vertex_size <- 5
> edge_width <- 0.2

# Plot the graph with undirected edges (without arrowheads)
> plot(G, layout = layout, vertex.color = colors[rep(1:num_communities, each = nodes_per_community)], vertex.size =
vertex_size, edge.color = "black", edge.alpha = 0.5, edge.width = edge_width, .label = NA, vertex.frame.color = "white", main
= "Random Clustered Network", arrow.mode = 0, edge.arrow.size = 0)

# Add a legend for community colors
> legend("topright", legend = paste("Community", 1:num_communities), fill = colors, bg = "white")
```

Output

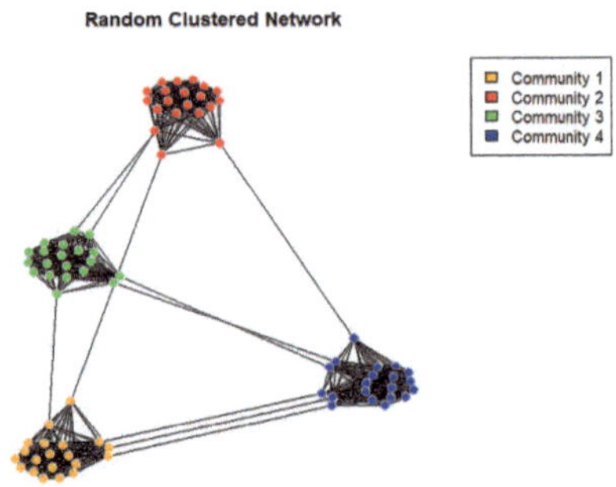

4.2.6 Python Code and Output

```
# Importing the required packages
>>> import igraph as ig
>>> import random
>>> import matplotlib.pyplot as plt
>>> import matplotlib.patches as mpatches

# Set the number of nodes per community and the number of communities
>>> nodes_per_community = 20
>>> num_communities = 4

# Create an empty graph
>>> G = ig.Graph(n=nodes_per_community * num_communities)

# Add nodes to the graph
>>> G.vs['name'] = [str(i) for i in range(1, nodes_per_community * num_communities + 1)]

# Add edges within communities
>>> for c in range(num_communities):
    community_nodes = list(range(c * nodes_per_community, (c + 1) * nodes_per_community))
    G.add_edges([(i, j) for i in community_nodes for j in community_nodes if i < j])

# Add a few intercommunity edges
>>> inter_edges = 10
>>> for _ in range(inter_edges):
    source_community = random.choice(range(1, num_communities + 1))
    target_community = random.choice(list(set(range(1, num_communities + 1)) - {source_community}))
    source_node = random.choice(list(range((source_community - 1) * nodes_per_community, source_community *
nodes_per_community)))
    target_node = random.choice(list(range((target_community - 1) * nodes_per_community, target_community *
nodes_per_community)))
    G.add_edge(source_node, target_node)

# Generate a layout for the graph
>>> layout = G.layout_fruchterman_reingold()

# Assign pastel colors to each community
>>> colors = ["#FFA500", "#FF0000", "#00FF00", "#0000FF"]

# Set the size of the nodes and width of the edges
>>> vertex_size = 0.3
>>> edge_width = 0.2

# Create a figure and plot the graph with undirected edges (without arrowheads)
>>> fig, ax = plt.subplots(figsize=(8, 6))
>>> ig.plot(G, layout=layout, vertex_color=[colors[i // nodes_per_community] for i in range(nodes_per_community *
num_communities)], vertex_size=vertex_size, edge_color="black", edge_opacity=0.5, edge_width=edge_width,
vertex_label=None, target=ax)

# Add a legend for community colors using matplotlib
>>> communities_legend = [f"Community {c}" for c in range(1, num_communities + 1)]
>>> legend_text = "\n".join(communities_legend)
>>> legend_patches = [mpatches.Patch(color=colors[i], label=communities_legend[i]) for i in range(num_communities)]

>>> x, y = 0.75, 1 # Adjust the coordinates (x, y) to change the position of the legend
>>> ax.legend(handles=legend_patches, bbox_to_anchor=(x, y), loc='upper left', facecolor='white', framealpha=0.8)

>>> plt.show()
```

Output

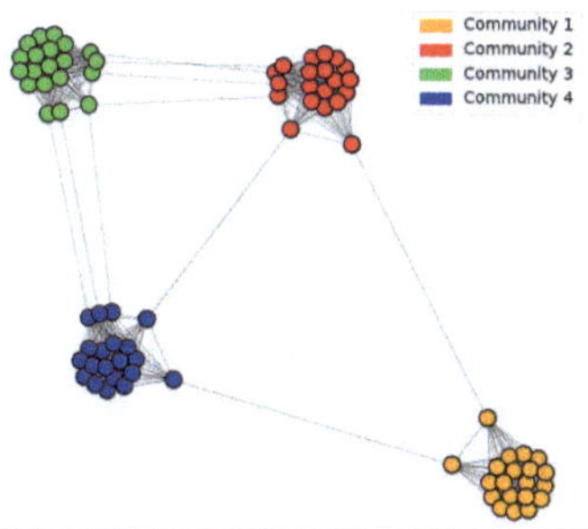

4.3 Core-Periphery Structure

Networks can be examined and analyzed from various perspectives, each offering unique insights into their underlying structures. These perspectives include the local, global, and meso (intermediate) levels. As specified previously, network theory aims to uncover statistical summaries for large networks, facilitating their comparison and exploration of their architectures. To achieve this, it is important to identify meso-level graph structures, which unveil significant features that remain hidden when solely examining individual nodes (or vertices) and edges (local level), or when summarizing the network as a whole (global level). One prominent meso-level structure is the community structure, characterized by cohesive groups of densely interconnected vertices known as communities. Within these communities, connections between vertices in different groups are relatively sparse. Extensive research on community structures has proven fruitful (Fortunato, 2010), yet other meso-level structures have not received the same level of attention. Among these, the core-periphery structure stands out. In core-periphery structures, networks consist of a core of vertices that are highly interconnected and a periphery of sparsely connected vertices. The core vertices are not only well connected to each other but also to the peripheral vertices, making them central players in the network's overall dynamics. Furthermore, the concept of meso-level structures is not limited to a single type of network. Instead, it is a flexible framework applicable to a wide array of domains. For instance, mesoscale structures have found relevance in disciplines like sociology, where social networks exhibit community-like patterns, and in international relations and economics, where core-periphery structures help model asymmetric power dynamics and trade relationships (Krugman, 1996). Over time, various methods and algorithms have been developed to detect and analyze meso-level structures. Community detection algorithms, such as modularity-based approaches and hierarchical clustering, have been instrumental in uncovering communities within networks. On the other hand, core-periphery detection methods, like the one proposed by Borgatti and Everett (2000), have shed light on the central role of core vertices in different network structures. The notion of "centrality" plays a crucial role in distinguishing core-periphery structures from community structures. Centrality metrics measure the importance of a vertex or network component, aiding in the identification and differentiation of these meso-level structures. Moreover, networks can exhibit nested core-periphery structures or a combination of core-periphery and community structures at the same time (Ahn et al., 2010; Fortunato & Hric, 2016). This complexity necessitates the development of advanced algorithms capable of investigating both types of mesoscale structures simultaneously. Hubs, vertices with a large number of connections (high degree), are common in real-world networks. Their presence can present challenges in community detection, as they often have strong ties to various parts of the network, making it challenging to assign them to specific communities accurately. In such cases, it may be more appropriate to consider hubs as part of the core in a core-periphery structure. Incorporating overlapping community methods can offer a more comprehensive understanding of the actual mesoscale network structure when hubs are involved.

The study of mesoscale structures will remain a fundamental aspect of network analysis in the coming years. By exploring the interconnections between local, global, and meso-level patterns, scientists can effectively address real-world challenges, optimize network performance, and attain valuable insights into complex interconnected systems across diverse social and economic contexts.

4.4 The Small-World Concept

The concept of a "Small World" emerged from the social sciences and highlights the interesting idea that individuals within social networks are typically connected to each other through a relatively small number of intermediate connections. This notion is often captured by the expression "six degrees of separation", although the number six should not be taken literally.

The small-world phenomenon was popularized by a famous experiment conducted by social psychologist Stanley Milgram in the 1960s (Milgram, 1967a, 1967b). Milgram asked participants to send letters to a specific target person, but they could only forward the letter to someone they knew personally. Surprisingly, the average number of intermediaries in the chain was approximately six, hence the term "six degrees of separation". His experiment revealed that, on average, people could be linked to one another through just a few intermediary connections, challenging the perception of a vast and disconnected social world. This finding underscored the remarkable interconnectedness of human social networks (Travers & Milgram, 1977). Beyond its origins in sociology, the concept of small-world networks has found applications across various fields. Small-world networks exhibit two key characteristics: a short average path length between vertices and a high level of clustering. The short path length implies that even in large networks, individuals can be reached through a relatively small number of steps. The high clustering signifies the presence of tightly-knit groups within the network, which enhances local interactions and facilitates information diffusion. While the term "Small World" is commonly used, it is important to note that it does not have a strict, quantifiable definition or a specific statistical test to determine if a given empirical network belongs to this class. The average distance between vertices in a network is influenced by its size, and larger networks tend to have greater distances between vertices. Therefore, the small-world property is often associated with network models where the average path length, denoted as d, increases at a slower rate or equal to the logarithm of the network size, expressed as $d \sim log(N)$, where N represents the network size and the logarithm scales as N approaches infinity. The Watts–Strogatz model, introduced by Duncan J. Watts and Steven H. Strogatz in 1998 (Watts & Strogatz, 1998), is a network model that aims to capture the small-world properties observed in real-world networks. It provides a way to generate networks with both high clustering and short average path lengths, two key characteristics of small-world networks. The small-world property is effectively captured by the Erdős–Rényi model we discussed earlier, but this model fails to adequately represent the local clustering observed in many real-world networks. Local clustering refers to the tendency of neighboring nodes to be connected to each other in tighter groups than would be expected by chance. This feature is important because it reflects the tendency of nodes to have connections with common neighbors, thus forming richer local connection structures than predicted by the Erdős–Rényi model. In social networks, which consist of mutual friendships or acquaintances, numerous studies have demonstrated significantly higher clustering coefficients compared to what the Erdős–Rényi model predicts for a corresponding random network. In contrast to the Erdős–Rényi model, the Watts–Strogatz model allows for the generation of networks with high local clustering, similar to what is observed in many real-world networks, while still maintaining short average path lengths. By starting with a regular ring lattice and introducing random rewiring of ties, the Watts–Strogatz model creates networks that exhibit a balance between local structure and long-range connections. This rewiring process enables the formation of shortcuts or long-distance connections, which contribute to reducing the average path length of the network (Watts & Strogatz, 1998). The construction of the Watts–Strogatz model can be summarized as follows:

1. Ring lattice creation: Begin with a ring of N nodes, with each node connected to its k closest neighbors. For instance, if k equals 2, each node is connected to the node on its immediate left and right, creating a simple circular chain of nodes.
2. Edge rewiring: Select a node and the connection to its nearest neighbor in a circular direction. With a probability of p, relocate this connection to a node randomly selected from all existing nodes in the ring. If not selected for rewiring, the connection remains as is. Continue with this process in a circular direction for all nodes in the ring, each time considering the link that connects a node to its nearest circular neighbor.
3. Process iteration: Once the first round of rewirings has been completed, start a new cycle. This time, for each node, consider the link that connects it to its second closest neighbor in a circular direction. With a probability p, follow the same rewiring process as in step 2.
4. Continual iteration: Keep repeating this iterative process, for each cycle considering the next farthest neighbor in the circular direction, until each original connection in the lattice has been evaluated for rewiring.

After this process, the resulting network constitutes a Watts–Strogatz small-world network. Here, p controls the level of randomness in the network. For p equals 0, the original ring lattice remains unchanged. For p equals 1, the network is completely random. For p values between 0 and 1, the network exhibits small-world properties.

The degree distribution in the Watts–Strogatz model provides insight into the number of connections for each individual node. When the rewiring probability, denoted as p, is set to zero, the model manifests as a ring structure where all nodes possess an equal number of connections. This configuration generates a harmonious community characterized by uniform connectivity, where each node shares an equivalent level of interaction. However, when p approaches one, making our model a random network, things change. The connections each node has start to vary, displaying an average behavior, a sort of independent randomness represented by a Poisson distribution. However, even among this randomness, the connections of a node never fall below $\frac{k}{2}$. Equation 4.7 describes the probability that a randomly selected node from the network has k connections.

$$P(k)=\begin{cases}0, & k \succ \langle k\rangle/2\\ \sum_{n=0}^{\min\left(k-\frac{\langle k\rangle}{2},\frac{\langle k\rangle}{2}\right)}\begin{pmatrix}\frac{\langle k\rangle}{2}\\ n\end{pmatrix}(1-p)^n p^{\frac{\langle k\rangle}{2}-n}\frac{\left(p\frac{\langle k\rangle}{2}\right)^{k-\frac{\langle k\rangle}{2}-n}}{\left(k-\frac{\langle k\rangle}{2}-n\right)!}e^{-p\frac{\langle k\rangle}{2}}, & k \succ \langle k\rangle/2\end{cases} \tag{4.7}$$

When $k < \langle k\rangle/2$, $P(k) = 0$, the Eq. (4.7) implies that no node has fewer connections than half the average degree of the network; when $k > \langle k\rangle/2$, $P(k)$ takes a more complex form. This part is based on the probability (p) of rewiring a node, the average degree $\langle k\rangle$, and the specific number of connections (k) of a node. This complex Equation (4.7) allows for the description of how the degree distribution changes from a delta function $P(k) = \delta(k - \langle k\rangle)$ for $p = 0$ to a Poisson distribution at $p = 1$ in the Watts–Strogatz model (Barrat & Weigt, 2000).

In their seminal work, Newman et al. (2000) calculated the average path length in the Watts–Strogatz model denoted by:

$$\begin{cases}\frac{N}{2\langle k\rangle}, & N\langle k\rangle p/2 \ll 1\\ \frac{\log\left(N\langle k\rangle p\right)}{\langle k\rangle^2 p}, & N\langle k\rangle p/2 \gg 1\end{cases} \tag{4.8}$$

The Eq. 4.8 suggests that a small-world property emerges when the number of rewired edges in the network exceeds a constant threshold denoted by $O(1)$, indicating constant time complexity. This means that an algorithm takes the same amount of time to complete its operations, regardless of the size of the input.

This small-world attribute essentially means that any two nodes within the network can be linked by a relatively short path, which allows for rapid information flows across the network. The clustering coefficient, denoted by C, is a crucial element of the Watts–Strogatz model. This coefficient quantifies the tendency for nodes to cluster together, creating tightly-knit communities within the larger network. When the rewiring probability (p) is zero, we are looking at a scenario where each node in the network is connected to an equal number of immediate neighbors, forming a circle or "ring". In this case, all nodes have an equal and high chance of forming clusters, leading to a high clustering coefficient. As the rewiring probability increases, the structure of the network changes. The nodes that were once only connected to their immediate neighbors in the ring may now find themselves connected to nodes further away. However, not all connections are rewired—some hold their ground, especially when the rewiring probability is not excessively high. This means that there is still a chance that some of the tightly-knit groups of nodes from the ring structure remain intact, although this likelihood decreases as the rewiring probability increases. This leads to an interesting outcome: as the rewiring probability increases, the clustering coefficient decreases, meaning that the network becomes less "exclusive". However, this happens at a slower step than the decrease in the average shortest path between nodes, a metric that evaluates the overall accessibility within the network. This dynamic results in a network that, within certain ranges of the rewiring probability, has both a small average path length (like a random network) and a high clustering coefficient (like a regular lattice). This unique combination of properties allows the Watts–Strogatz model to represent the small-world phenomenon seen in many real networks.

4.5 Barabási–Albert (BA) Model

Power-law distribution is surprisingly common in various phenomena, and they often appear in real-world networks. Many such instances of power-law distribution can be found, signifying the existence of scale-free networks (Barabási & Albert, 1999; Albert & Barabási, 2002). The power-law degree distribution in these networks is primarily due to two universal mechanisms:

- Growth: Networks are constantly evolving as new vertices are added over time.
- Preferential attachment: New vertices are more likely to connect to vertices that already have a large number of connections.

A model incorporating these two factors can successfully generate scale-free distributions, as observed in reality. These concepts were first proposed in a pivotal paper by Barabási and Albert (1999). The Barabási–Albert (BA) model is a specific type of preferential attachment model. In other words, the BA model is a way to implement the concept of preferential attachment. The Barabási–Albert (BA) model uses the principle of preferential attachment to generate a random scale-free network. Real-world networks are dynamic, meaning they can accommodate new vertices over time. Hence, the set of vertices *(V)* is dynamic and grows in size. In the Barabási–Albert model, the size of the network increases linearly over time, with the rate of addition being a constant equal to 1. The principle of preferential attachment can be thought of as the "rich get richer" mechanism in the context of network growth. It is based on the idea that popular vertices are more likely to attract new connections. For instance, in a social network where vertices represent people and edges represent friendships, a new person joining the network is more likely to become friends with those who already have many friends. This is because people with many friends are often seen as more sociable, trustworthy, or influential. Thus, vertices that acquire more connections early on will tend to gain connections at a faster rate — a classic "rich get richer" scenario. This phenomenon is the core of the preferential attachment mechanism. The Barabási–Albert model quantifies this by defining the probability that the new vertex will connect to an existing vertex as proportional to the number of existing connections that vertex already has. So, vertices with more connections have a higher chance of being chosen. This can be mathematically represented as:

$$P(k_i) = \frac{k_i}{\sum_{\forall j} k_j} \tag{4.9}$$

Here, $P(k_i)$ is the probability of the new vertex connecting to the existing vertex i, where k_i is the degree of the existing vertex i, and the denominator is the sum of degree of all vertices in the graph. Unlike in the E-R random network where the connection probability is a fixed value, in the Barabási–Albert model, the probability is not a constant for all vertices in the network. Considering the Barabási–Albert model as a framework for comprehending urban growth, a city starts with a small population (the initial vertices). Subsequently, as time progresses, more people move to the city, and new neighborhoods form (new vertices are added). As time goes on, new neighborhoods are established. Each of these new neighborhoods does not exist in isolation—they build connections to existing parts of the city, like roads or public transit links. In the Barabási–Albert model, these connections are the m edges that each new vertex adds (growth mechanism). With every time step, a new neighborhood is established, bringing along m new connections to the existing parts of the city. Now, not all neighborhoods are equally likely to get a new connection. The more connected a neighborhood already is, the more likely it is to get even more connections. This is because well-connected neighborhoods are more accessible and attractive (preferential attachment mechanism). After t time steps, the city grows to include $m_o + t$ neighborhoods (vertices) and mt connections (edges). As the city grows over time, neighborhoods' popularity (the number of connections they have) follows a power-law distribution, meaning that while most neighborhoods have a few connections, a few neighborhoods become very popular, having many more connections than the average. Both growth (the addition of new neighborhoods) and preferential attachment (the tendency to connect more to popular neighborhoods) are necessary for the city's development to reflect the distribution observed in real-world networks. If either of these aspects is overlooked, the result would not be a city with a few highly popular neighborhoods and many less popular ones, which is characteristic of scale-free networks (Cordoba, 2008).

4.5.1 *Properties of the Barabási–Albert (BA) Model*

The dynamic properties of the Barabási–Albert (BA) model are explored through a variety of analytical techniques. Here, we analyze one of these techniques, namely the master equation, to study the probability distribution in the Barabási–Albert (BA) model. This approach provides a mathematical description of how the network changes over time. In the BA model, the master equation helps in characterizing the dynamic properties of the model, specifically the probabilities of vertices in a network having certain degrees at given points in time. The function *p(ki,* v_i*, t)* symbolizes the probability of a vertex *v* possessing a degree *k* at a specific point in time, denoted as *t*. Consequently, *p(k, t)* is representative of the comprehensive degree distribution of the network at time *t*. On the other hand, *p(k)* which equals *p(k,* $t \rightarrow \infty$*)* represents a stationary degree distribution of the network, a distribution that does not vary over time (Dorogovtsev et al., 2000; Dorogovtsev & Mendes, 2001).

Moving forward, the degree probabilities for an individual vertex can be depicted with the following master equation:

$$P(k,v_i,t+1) = \frac{k-1}{\sum_{j\leq t} k_j} P(k-1,v_i,t) + \left(1 - \frac{k}{\sum_{j\leq t} k_j}\right) P(k,v_i,t) \quad (4.10)$$

The first term in this Eq. (4.10) refers to the scenario where a new edge is connected to the vertex v_i when its degree is *(k − 1)* at time *t*. The second term hints at a situation where the degree *k* of vertices v_i does not increase at time *t*. Notably, the denominator $\sum_{j \leq t} k_j$ refers to a sum of the degrees of all vertices in the network at time *t*. In the BA model context, when *m* is 1 (which refers to the number of edges added at each time step), this means that each new vertex introduced to the network forms an edge with an existing vertex. Therefore, the total number of edge endpoints, or total degrees of all vertices, grows linearly with time. Each vertex's degree is the number of edges (or connections) it has.

Further manipulation of equation in 4.10 provides us with:

$$2t\left[P(k,v_i,t+1) - P(k,v_i,t)\right] = (k-1)P(k-1,v_i,t) - dP(k,v_i,t) \quad (4.11)$$

When we shift to the continuous limit in both time *t* and degree *k*, we achieve the equation:

$$2t\frac{\partial P(k,v_i,t)}{\partial t} + \frac{\partial\left[dP(k,v_i,t)\right]}{\partial t} = 0 \quad (4.12)$$

Moreover, the average connectivity of a single vertex v_i at time *t* is calculated as:

$$\bar{k}(v_i,t) = \sum_{k=1}^{\infty} k \cdot P(k,v_i,t) = \int_0^{\infty} k \cdot P(k,v_i,t)\,dk \quad (4.13)$$

By applying the integral operation to the above differential Eq. (4.13) and integrating its right side, we get:

$$\frac{\partial \bar{k}(v_i,t)}{\partial t} = \frac{\bar{k}(v_i,t)}{\int_0^{t} \bar{k}(v,t)\,dv} = 0 \quad (4.14)$$

The core idea of the Eq. 4.14 is quite simple. Each new edge is uniformly distributed among all vertices, taking the preferences into account. It refers to the process of how new connections form in the network (uniformly because each vertex has an equal opportunity to be chosen for the formation of a new link), considering the mechanism of preferential attachment. This modeling technique enables us to analytically determine the degree distribution of a network (Dorogovtsev et al., 2000; Dorogovtsev & Mendes, 2001; Jeong et al., 2003).

Assuming that both $P(k)$ and $k(v, t)$ exhibit scaling behavior, we can say $P(k)$ equals $k^{-\alpha}$, and $k(v, t)$ equals v^{γ}. Subsequently, we assume the scaling relationship α equals $1 + 1/\theta$, which indicates a direct relation between the degree distribution exponent α and the temporal growth exponent γ, offering insights into the system's global properties. These conclusions denote that networks generated by the BA model have the scale-free property. Scale-free networks are characterized by a power-law degree distribution, which implies that while most vertices have a few connections, there are a significant number of hubs with a very high degree.

We have seen that the Barabási–Albert (BA) model explains how networks grow and evolve over time. One of the key principles it uses is known as the "continuum theory". This theory mainly looks at how a vertex in a network builds connections as time evolves. In simple terms, the BA model suggests that vertices with more connections are likely to gain even more over time. Think of it as a scientific context, where the more popular scientists tend to attract more attention, hence more connections. This model shows that small initial differences in connections between nodes can lead to big differences over time as the network grows. Another important part of the BA model is understanding how often nodes have a certain number of connections. In networks generated by the BA model, we have observed that the distribution of connections among nodes follows a power-law distribution. Above we addressed that this pattern reveals that the network has a "scale-free" nature, characterized by few highly connected nodes and many nodes with few connections. When we examine how closely nodes in a network group together, using a measure known as the clustering coefficient, we find that networks generated through the BA model behave differently from random networks. As the network size grows, the clustering in the BA model decreases more slowly than in random networks. However, unlike many real-world networks, the average clustering in the BA model does reduce as the network size grows. Another crucial aspect of networks built using a Barabási–Albert model is the average path length, i.e., the number of links required to connect any two nodes (Albert & Barabási, 2000; Jeong et al., 2003). The BA model shows a small-world effect, where the average path length increases slowly as the network grows. This behavior is similar to many real-world networks (Barabási et al., 2002).

Finally, the BA model reveals that there is a complex relationship between the number of connections that linked nodes have. This relationship suggests that the number of connections a node has depends not only on its own characteristics but also on the characteristics of the nodes it is connected to. This interconnected relationship is one of the key differences between the BA model and random networks.

4.5.2 Stata Code and Output

```
# Clear any existing network data
. nwclear

# Generating a small-world network with 40 nodes, where each node is connected to 2 other nodes (k(2)), and the probability of each connection is 0.2.
. nwsmall 40, k(2) prob(.2)

# Plotting the network according to a multidimensional scaling (MDS) algorithm
. nwplot, layout(mds)
```

Output

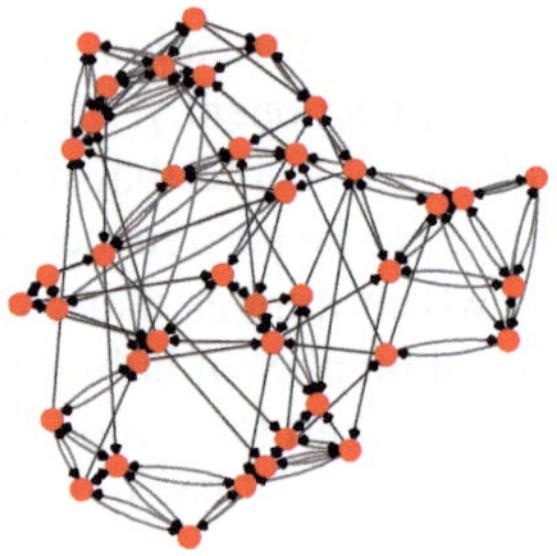

```
# Clear any existing network data
. nwclear

# Generating an undirected scale-free network of 50 nodes with 3 connected nodes at start and 3 connections each new node forms
. nwpref 20, undirected m0(3) m(3)

# Plotting the network according to a multidimensional scaling (MDS) algorithm
. nwplot, layout(mds)
```

Output

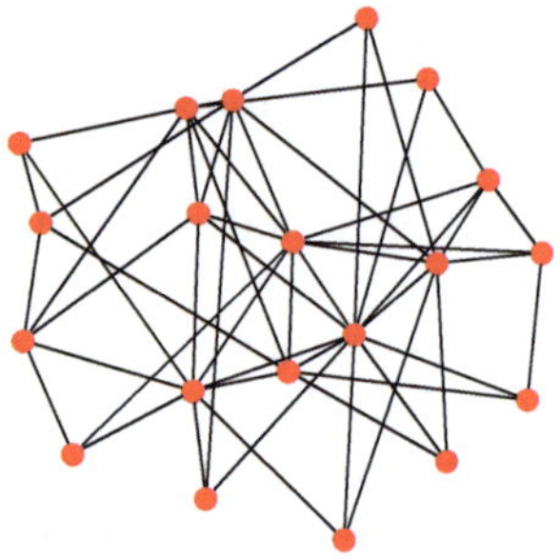

4.5.3 R Code and Output

```
# Create matrices to store average path length (L) and clustering coefficient (C) values
> L <- matrix(nrow=25, ncol=6)
> C <- matrix(nrow=25, ncol=6)

# Define the rewiring probabilities
> s <- c(0.00001, 0.0001, 0.001, 0.01, 0.1, 1)

# Create a lattice graph with 500 nodes and neighbors connected up to 10 neighbors
> lat <- make_lattice(dimvector=500, nei=10, circular=TRUE)

# Loop through different rewiring probabilities and calculate L and C for 25 replicates
> for (j in 1:6) {
  for (k in 1:25) {
   # Rewire the lattice graph with the current rewiring probability
   g <- rewire(lat, with=each_edge(prob=s[j], loops=FALSE))

# Calculate the average path length and clustering coefficient for the rewired graph
   L[k, j] <- average.path.length(g)
   C[k, j] <- transitivity(g, "localaverage")
  }
}

# Calculate the average path length and clustering coefficient of the original lattice graph
> L0 <- average.path.length(lat)
> C0 <- transitivity(lat, "localaverage")

# Calculate the relative values of L and C for each rewiring probability
> Lp <- L / L0
> Cp <- C / C0

# Calculate the mean values of Lp and Cp for each rewiring probability
> mean_Lp <- colMeans(Lp)
> mean_Cp <- colMeans(Cp)

# Plot the mean values of Lp and Cp
> plot(s, mean_Lp, type="b", pch=20, log="x", ylim=c(0, 1), xlab="p", ylab="")
points(s, mean_Cp, type="b", pch=0, ylim=c(0, 1), xaxt="n", yaxt="n", xlab="", ylab="")
legend("bottomleft", pch=c(20, 0), legend=c("Average Path Length (Lp)", "Clustering Coefficient (Cp)"))

# Create Erdös-Rényi random graph (g1) and Barabási-Albert preferential attachment model (b1)
> g1 <- erdos.renyi.game(50, p=0.15)
> b1 <- barabasi.game(50, power=1, m=3, directed=FALSE)

# Plot the graphs side-by-side with proportional vertex sizes
> par(mfrow=c(1, 2), mar=c(1, 1, 3, 1))
> plot(g1, layout=layout.fruchterman.reingold, vertex.size=degree(g1), vertex.label="", vertex.color="black", main="Erdos-
Renyi Random Graph")
> plot(b1, layout=layout.fruchterman.reingold, vertex.size=degree(b1), vertex.label="", vertex.color="black",
main = "Preferential Attachment")
```

Output

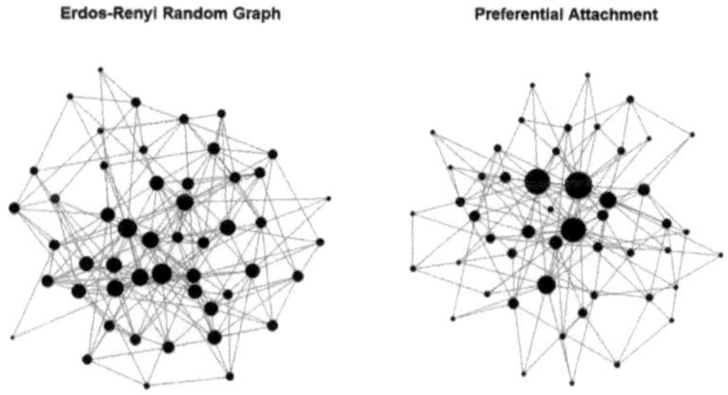


```
# Creating the graphs
> g <- erdos.renyi.game(4000, p = 0.002)
> b <- barabasi.game(4000, power = 1, m = 5, directed = FALSE)

# Summarizing the graphs
> summary(g) # Prints a summary of the Erdős–Rényi graph 'g'
> summary(b) # Prints a summary of the Barabási graph 'b'
```

```
# Calculating degree distributions
> deg.dist.g <- degree.distribution(g) # Degree distribution of graph 'g'
> deg.dist.b <- degree.distribution(b) # Degree distribution of graph 'b'

# Plotting the degree distributions
> par(mfrow = c(1, 2)) # Set up a 1x2 grid for side-by-side plots

# Plot the degree distribution of the Erdős–Rényi random graph 'g' using log-log scale
> plot(deg.dist.g, pch = 20, xlab = "k", ylab = "P(k)", las = 1, main = "Erdos-Renyi Random Graph", log = "xy", col = "red", cex = 1.5)

# Plot the degree distribution of the Barabási graph 'b' using log-log scale
> plot(deg.dist.b, pch = 20, xlab = "k", ylab = "P(k)", las = 1, main = "Preferential Attachment", log = "xy", col = "red", cex = 1.5)
```

Output

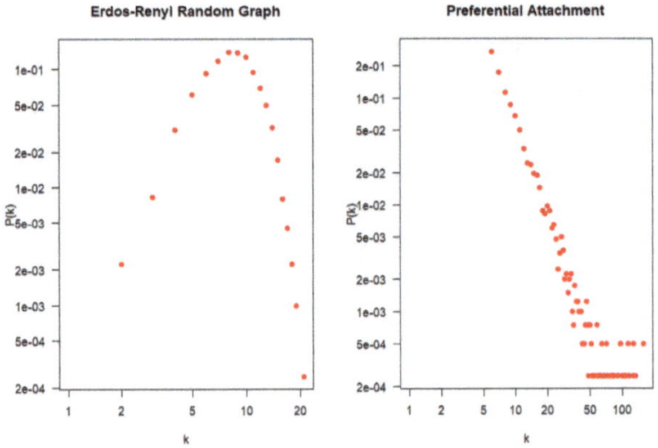

4.5.4 Python Code and Output

```
# Run this line if you have not installed these packages yet:
# !pip install networkx numpy matplotlib

>>> import networkx as nx
>>> import numpy as np
>>> import matplotlib.pyplot as plt

# Initialize matrices to store average path length (L) and clustering coefficient (C) values
>>> L = np.zeros((25, 6))
>>> C = np.zeros((25, 6))

# Define the rewiring probabilities
>>> s = [0.00001, 0.0001, 0.001, 0.01, 0.1, 1]

# Create a lattice graph with 500 nodes and neighbors connected up to 10 neighbors
>>> lat = nx.watts_strogatz_graph(500, 20, 0)

# Loop through different rewiring probabilities and calculate L and C for 25 replicates
>>> for j in range(6):
  >>> for k in range(25):
    # Rewire the lattice graph with the current rewiring probability
    >>> g = nx.watts_strogatz_graph(500, 20, s[j])

    # Calculate the average path length and clustering coefficient for the rewired graph
    >>> L[k, j] = nx.average_shortest_path_length(g)
    >>> C[k, j] = nx.average_clustering(g)

# Calculate the average path length and clustering coefficient of the original lattice graph
>>> L0 = nx.average_shortest_path_length(lat)
>>> C0 = nx.average_clustering(lat)

# Calculate the relative values of L and C for each rewiring probability
>>> Lp = L / L0
>>> Cp = C / C0

# Calculate the mean values of Lp and Cp for each rewiring probability
>>> mean_Lp = np.mean(Lp, axis=0)
>>> mean_Cp = np.mean(Cp, axis=0)

# Plot the mean values of Lp and Cp
>>> plt.figure()
>>> plt.loglog(s, mean_Lp, marker='o', linestyle='-', label='Average Path Length (Lp)')
>>> plt.loglog(s, mean_Cp, marker='o', linestyle='-', label='Clustering Coefficient (Cp)')
>>> plt.xlabel('p')
>>> plt.legend(loc='lower left')
>>> plt.show()
```

Output

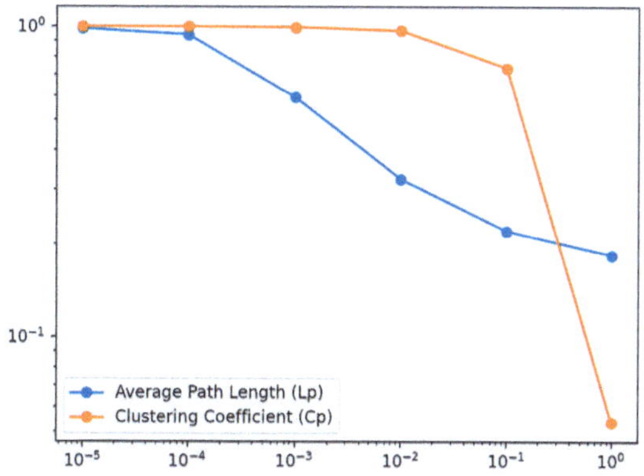

```
# Generate Erdős–Rényi random graph (g1) and Barabási-Albert preferential attachment model (b1)
>>> g1 = nx.erdos_renyi_graph(50, 0.15)
>>> b1 = nx.barabasi_albert_graph(50, 3)

# Plot the graphs side-by-side with proportional vertex sizes
>>> fig, axes = plt.subplots(1, 2, figsize=(12, 6))

# Draw Erdős–Rényi random graph
>>> nx.draw(g1, ax=axes[0], node_size=[deg * 10 for _, deg in g1.degree()], node_color='black', with_labels=False)
>>> axes[0].set_title('Erdős–Rényi Random Graph')

# Draw Barabási-Albert preferential attachment model
>>> nx.draw(b1, ax=axes[1], node_size=[deg * 10 for _, deg in b1.degree()], node_color='black', with_labels=False)
>>> axes[1].set_title('Preferential Attachment')

>>> plt.show()

# Output the degree distributions
>>> degree_sequence_g1 = [deg for _, deg in g1.degree()]
>>> degree_sequence_b1 = [deg for _, deg in b1.degree()]

>>> plt.figure(figsize=(10, 6))
>>> plt.loglog(range(1, len(degree_sequence_g1) + 1), sorted(degree_sequence_g1, reverse=True), 'r.', label='Erdős–Rényi')
>>> plt.loglog(range(1, len(degree_sequence_b1) + 1), sorted(degree_sequence_b1, reverse=True), 'b.', label='Preferential Attachment')
>>> plt.xlabel('Rank')
>>> plt.ylabel('Degree')
>>> plt.title('Degree Distributions of Erdős–Rényi and Preferential Attachment Graphs')
>>> plt.legend()
>>> plt.show()
```

Output

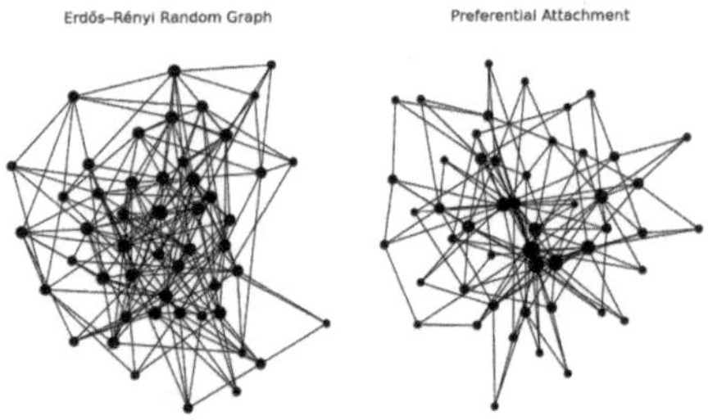


```
# Output the degree distributions
>>> degree_sequence_g1 = [deg for _, deg in g1.degree()]
>>> degree_sequence_b1 = [deg for _, deg in b1.degree()]

>>> plt.figure(figsize=(10, 6))
>>> plt.loglog(range(1, len(degree_sequence_g1) + 1), sorted(degree_sequence_g1, reverse=True), 'r.', label='Erdős–Rényi')
>>> plt.loglog(range(1, len(degree_sequence_b1) + 1), sorted(degree_sequence_b1, reverse=True), 'b.', label='Preferential Attachment')
>>> plt.xlabel('Rank')
>>> plt.ylabel('Degree')
>>> plt.title('Degree Distributions of Erdős–Rényi and Preferential Attachment Graphs')
>>> plt.legend()
>>> plt.show()
```

Output

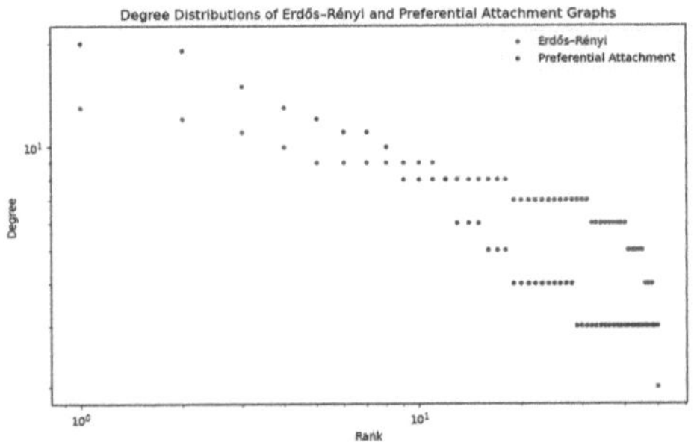

4.6 Power-Law: A Formal Specification

Many real-world data demonstrate a tendency to group around an average value. This is observable in a variety of situations, such as the housing prices, the heights of particular tree species in a forest, or the interest rates in the banking sector. In these examples, instances that significantly deviate from the typical value are given a very low probability, thereby making the average value a fair representative of the majority of measurements. This characteristic aligns with the principles outlined in the central limit theorem. Nevertheless, this characteristic is not universally applicable to all distributions. In particular scenarios, deviations from this pattern serve as indicators of complexities in the underlying processes governing these distributions. Over the past 10 years, a vast number of distributions that deviate from the Gaussian distribution have been unearthed in complex systems. These span a diverse range of disciplines, prominently including the social and economic sciences. A power-law distribution is a unique type of statistical distribution, distinctly differing from the Gaussian or normal distribution, as it lacks a typical average value.

A power-law represents a type of relationship between two quantities, where one quantity varies as a power of the other, and this relationship exhibits scale invariance. Scale invariance is a feature of certain mathematical functions that stay the same even when you change the size (or "scale") of specific entities (Clauset et al., 2009). An example from the physical world is a fractal. The term "scale invariance" refers to the idea that the fundamental pattern or behavior remains unchanged (invariant) when the scale of the system is altered. Power-law distributions are typically used to establish a relationship between two variables. This relationship can generally be expressed in the form of:

$$f(x) = cx^{-\alpha} + z(x^{-\alpha}) \tag{4.15}$$

where k and α are constants, and $z(x^{-\alpha})$ is a function of x^{α} that approaches zero as x becomes large.

In this context, α is often labeled as the scaling exponent. The name "scaling" represents the characteristic of a power-law function to adhere to the equation:

$$f(cx) \propto f(x) \tag{4.16}$$

where c is a constant, signifying that the function maintains its form during scale changes. Therefore, by rescaling the argument of the function, a change happens to the constant of proportionality, but the original form of the function still remains. This notion becomes more comprehensible when applying a logarithm to each side of the equation:

$$\log(f(x)) = -\alpha \log x + \log c \tag{4.17}$$

Note that this Eq. (4.17) takes on the form of a linear relationship characterized by a slope of α. When the argument is rescaled, it leads to a vertical linear shift in the function, yet its fundamental shape and slope stay unaltered. This implies a uniform increase or decrease in the values produced by the function, like to shifting the function vertically on a graph. However, it remains critical to understand that such alterations neither influence the inherent structure of the function nor cause changes to the slope (α). In the logarithmic form of the function $\log(f(x)) = -\alpha \log x + \log c$, the slope of the function is represented by α. This follows from the standard form of a linear equation, $y = mx + b$, where m is the slope of the line (Clauset et al., 2009).

4.6.1 Characteristics of Power-Law Distribution

One of the appealing features of power-law that intrigue scientists is their inherent scale invariance. Given a relationship such as $f(x) = cx^{-\alpha}$, if the argument x is multiplied by a constant, it leads to the following equivalent scaling in the output of the function:

$$f(cx) = (cx)^{-\alpha} = c^{-\alpha} f(x) \propto f(x) \tag{4.18}$$

This indicates that any scaling by a constant just multiplies the original power-law relationship by the constant $c^{-\alpha}$. Consequently, all power-law models having a specific scaling exponent can be seen as having similar scale-invariant properties, apart from multiplicative constant factors. This feature is responsible for the linear relationship seen when taking the logarithms of both $f(x)$ and x, leading to a straight line on the log–log plot which is often associated with a signature of a power-law (Newman, 2005). However, it is worth noting that with real-world data, such linearity is necessary but not always sufficient for the data to follow a power-law relationship. There exist various methods to generate finite datasets that mimic this signature behavior, but they do not conform to a true power-law distribution in their asymptotic limit. As such, the accurate fitting and validation of power-law models continue to be a field of active research in statistics. In their seminal paper, Clauset et al. (2009) provide a thorough investigation into the process of fitting and validating power-law models within empirical datasets. The authors highlight that while many systems produce datasets that display a striking linear trend on a log–log plot — a feature often associated with power-law—not all these systems strictly adhere to power-law, particularly in their asymptotic limits. They propose robust methods for identifying and characterizing power-law in empirical data, including maximum likelihood fitting methods and statistical tests for comparing the goodness-of-fit of power-law models against alternative hypotheses. They also introduce a statistical framework for discerning if observed data can be modeled by a power-law or if another explanation is more plausible. This research has significantly contributed to the on-going understanding of power-law phenomena across a wide range of fields, reinforcing the importance of rigorous statistical methods in the analysis of empirical data. It underscores that, while power-law is an important tool in understanding complex systems, their identification and application require careful consideration to avoid misinterpretation of data patterns.

The association of power-law with a distinct scaling may indicate deeper dynamics that gave rise to the power-law relationship. In the social sciences, for instance, phenomena such as the distribution of wealth or city sizes often show power-law distributions. Often referred to as scaling exponents, the exponents of these distributions play a vital role in understanding the behavior of the system. Different socio-economic systems exhibiting the same scaling exponents—demonstrating similar scaling behavior under transformation—suggesting underlying commonalities (Cordoba, 2008).

For example, consider the distribution of income across different societies. Regardless of the specifics of each society, the income distribution often falls into a power-law pattern, especially in the tail representing the wealthiest individuals. Despite the diverse cultural, economic, and political differences, these societies exhibit a commonality in how wealth is distributed among their population, illustrating a kind of universality. This suggests that a majority of nodes have minimal connections, while a handful of nodes possess an exceptionally high number of connections. In these contexts, the scaling exponent provides valuable insights, contributing significantly to understanding the structure of the network as well as its resilience, dynamics, and functionality.

This pattern, although less exhaustively studied, has been identified in various self-organized systems, where the state of the system behaves as an attractor (Gabaix, 2009; Clauset et al., 2009; Lin et al., 2023). This shared feature, commonly termed universality, places systems with identical scaling exponents into the same universality class. Moreover, understanding this concept of universality can shed light on the temporal evolution of the system, its reaction to modifications, and the possibility of the emergence of new properties, thereby cementing its importance in the domain of Network Science and overarching system analysis.

4.6.2 Power-Law Distribution for Continuous and Discrete Data

In mathematics, a power-law distribution is a statistical pattern that describes how a particular quantity, e.g., x behaves in relation to its probability of occurrence. The key characteristic of a power-law distribution is that the probability of observing a value of x is proportional to a power of that value raised to a constant exponent. In other words, the likelihood of finding a specific value of x follows a relationship of the form:

$$p(x) \propto cx^{-\alpha} \tag{4.19}$$

Even though there are anomalies, it is common to observe that the parameter α (known as the "exponential parameter") of a power-law distribution falls within the range of $2 < \alpha < 3$. This parameter plays a crucial role in shaping the distribution curve (Albert & Barabási, 2002; Clauset et al., 2009).

It is important to note that the empirical data may show power-law-like behavior only within a specific range or beyond a certain threshold value. In power-law distribution, there is often a lower bound or minimum value, denoted as *x-min* below which the power-law behavior may not hold. However, for values greater than *x-min*, the tail of the distribution exhibits the characteristic power-law relationship. Power-law distribution can manifest in two main formats:

- Continuous data: This format involves real and continuous numbers. It implies that x can take on any real value within a specific range.
- Discrete data: In this case, x can only assume certain distinct values, usually positive integers.

Supposing that x means the quantity for which we are seeking the distribution, a continuous power-law distribution can be characterized by a probability density function, $p(x)$, as follows:

$$p(x)dx = \Pr(x \le X < x + dx) = cx^{-\alpha}dx \tag{4.20}$$

In this context, x represents the observed value and c is the normalization constant. This probability density function diverges when x approaches 0. The equation does not hold for all $x \ge 0x$; therefore, there must be a lower limit for the power-law distribution, which is the *x-min*.

Thus, assuming $\alpha > 1$, the normalized constant can be calculated as follows:

$$p(x) = \frac{\alpha - 1}{x_{\min}} \left(\frac{x}{x_{\min}} \right)^{-\alpha} \tag{4.21}$$

The function (4.21) describes the probability density of a continuous power-law distribution. Here, $p(x)$ indicates the probability of observing the value x. The α is the shape parameter, and *x-min* is the minimum value to which the power-law applies. Considering only integer values with a probability distribution that takes the form:

$$p(x) = \Pr(X = x) = cx^{-\alpha} \tag{4.22}$$

the distribution derived from 4.22 tends to infinity when x approaches 0. Therefore, there must exist a lower limit $x_{min} > 0$ in the power-law relationship. Calculating the normalized constant, we get:

$$p(x) = \frac{x^{-\alpha}}{\zeta(\alpha, x_{\min})} \quad \text{with} \quad \zeta(\alpha, x_{\min}) = \sum_{n=0}^{\infty} (n + x_{\min})^{-\alpha} \tag{4.23}$$

The denominator in the first part of Eq. (4.23) is known as the Hurwitz zeta function, which is a type of generalized function. This function plays a crucial role in the equation, normalizing the distribution to ensure that the total probability across all possible values is equal to 1, a fundamental requirement for any probability distribution. The Hurwitz zeta function is defined as:

$$\zeta(\alpha,s) = \sum_{x=0}^{\infty} \frac{1}{(n+\alpha)^s} \tag{4.24}$$

In many scenarios, it is quite beneficial to also examine the complementary cumulative distribution function (CCDF), denoted as $P(x)$, of a variable that follows a power-law distribution. Both in the continuous and discrete cases, the CCDF is defined as $P(x) = Pr(X \geq x)$, which means it shows the probability that the random variable X takes on a value greater than or equal to x. Examining the CCDF is especially useful in empirical data analysis for power-law because it can sometimes be more robust against fluctuations in the data. It is also a more natural way to present data that are subject to a "survival bias" where larger values are systematically more likely to be observed.

In the context of a continuous power-law distribution, the CCDF can be expressed as:

$$P(x) = \int_x^{\infty} p(x')\,dx' = \left(\frac{x}{x_{\min}}\right)^{-\alpha+1} \tag{4.25}$$

The expression in (4.25) is the integral of the power-law probability density function, $p(x)$, from x to infinity. The result, $P(x)$, describes the tail distribution, i.e., the probability that X is greater than or equal to x. This tail behavior, often heavy or fat, is one of the distinctive features of power-law distributions.

To solve this integral, it is easier to change variables. Let $\tau = \frac{x_0}{x_{\min}}$. Then, $dx_0 = x_{min}\, d\tau$ and the limits of the integral change to $\tau = \frac{x}{x_{\min}}$ and $\tau = \infty$:

$$\mathrm{P}(\mathrm{x}) = (\alpha - 1) \int_{x/x_{\min}}^{\infty} \tau^{-\alpha} x_{\min}\, d\tau \tag{4.26}$$

Simplifying gives:

$$\mathrm{P}(\mathrm{x}) = (\alpha - 1)\, x_{\min} \int_{x/x_{\min}}^{\infty} \tau^{-\alpha}\, d\tau \tag{4.27}$$

Finally, the integral of $\tau^{-\alpha}$ from x/x_{min} to infinity is equal to $\frac{\tau^{-\alpha+1}}{-\alpha+1}$, which gives the final result:

$$P(x)=\left(\frac{x}{x_{\min}}\right)^{-\alpha+1} \tag{4.28}$$

For the case of discrete variables, x can assume only a specific set of values. In this session, we restrict our attention to the scenario where these values are integers, with a probability distribution denoted by the equation in 4.19. As mentioned, this distribution approaches infinity as x approaches zero, which necessitates the existence of a lower bound, x_{min}, for the power-law behavior. Once we compute the normalization constant, we find:

$$P(x)=\frac{x^{-\alpha}}{\zeta(\alpha,x_{\min})} \tag{4.29}$$

with the denominator that represents the generalized or Hurwitz zeta function:

$$\zeta(\alpha,x_{\min})=\sum_{n=0}^{\infty}(n+x_{\min})^{-\alpha} \tag{4.30}$$

In many scenarios, it is beneficial to examine the complementary cumulative distribution function (CDF) for variables distributed according to a power-law. In the case of a discrete power-law distribution, the complementary cumulative distribution function (CCDF) is represented as:

$$P(x)=\frac{\zeta(\alpha,x)}{\zeta(\alpha,x_{\min})} \tag{4.31}$$

In numerous applications, continuous power-law behavior is often used to approximate its discrete counterpart for mathematical simplicity. However, one should proceed with caution as there are multiple methods to accomplish this approximation, and not all of them yield reliable results. A relatively reliable approach treats an integer power-law as if the values of x were derived from a continuous power-law and then rounded to the nearest integer. This technique tends to provide reasonably accurate outcomes in various applications. However, other methods, such as truncation or assuming proportionality between the probabilities of generating integer values in discrete and continuous scenarios, tend to produce less accurate results and should be avoided (Clauset et al., 2009).

4.6.3 R Code and Output

```
# Importing the required libraries
> library(ggplot2)
> library(poweRlaw)
> library(actuar)

# Set the alpha value for power-law
> alpha = 1.8
> nr = unique(round(10^(seq(1, 6, length.out = 40)))) # sizes of samples to simulate

# Initialize vectors to store sample means and variances
> mn = numeric(length(nr))
> vr = numeric(length(nr))

# Generate iid PL random variables, and compute the sample mean and variance
> for (i in seq_along(nr)) {
n = nr[i]
x = rpareto(n, scale = 1, shape = alpha)
mn[i] = mean(x)
vr[i] = var(x)
}

# Initialize the vectors for storing the sample means and variances
> mn <- rep(0, length(nr))
> vr <- rep(0, length(nr))

# Generate iid PL random variables, and compute the sample mean and variance
> for (i in seq_along(nr)) {
n = nr[i]
x = rpareto(n, scale = 1, shape = alpha) # Generate iid PL r.v. using rpareto function from actuar package
mn[i] = mean(x) # compute the sample mean
vr[i] = var(x) # compute the sample variance
}


# Create a dataframe for plotting
> df <- data.frame(Size=rep(nr, 2), Value=c(mn, vr), Type=rep(c("Sample Mean", "Sample Variance"), each=length(nr)))

# Plot
> ggplot(df, aes(x=Size, y=Value, color=Type)) +
geom_point(size=3) +
scale_x_log10() + scale_y_log10(limits = c(1, 10^3)) +    # setting y-axis limits
labs(x="Size of sample, n", y="Value") +
theme_minimal() +
# removing major grid lines
theme(plot.background = element_rect(fill="white", colour="black"), panel.grid.major = element_blank(),
panel.grid.minor = element_blank(),              # removing minor grid lines
axis.line = element_line(colour = "black")) +    # adding axis lines
annotate("text", x=8*10^4, y=10^3, label=paste("alpha =", alpha)) +
theme(legend.position="top") +
scale_color_manual(values=c("orangered", "blue1"))
```

Output

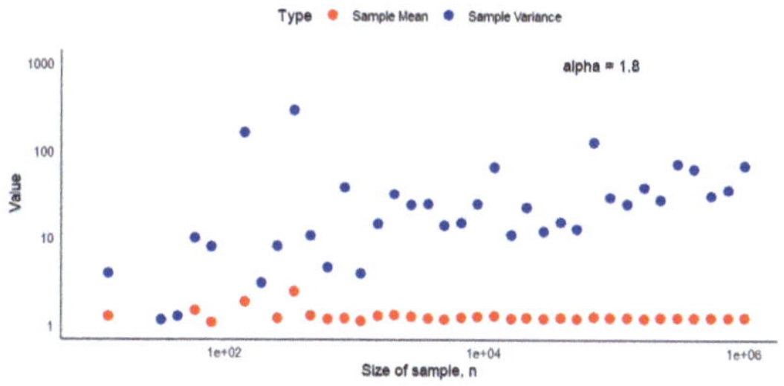

The plot in the output represents the mean values and variances of samples of random variables distributed according to a power-law. On the *x-axis*, we have the sample sizes, while the *y-axis* represents both the sample mean and variance. Each point represents a sample, with the color differentiating mean values (red) from variances (blue). The legend indicates which quantities the different colors correspond to.

4.6.4 Python Code and Output

```
>>> import numpy as np
>>> import powerlaw
>>> import matplotlib.pyplot as plt
>>> from time import sleep

# Set the alpha value for power-law
>>> alpha = 1.8

# Define the sample sizes to simulate
>>> nr = np.unique(np.round(np.logspace(1.3, 4.5, 40)))

# Initialize storage for sample means and variances
>>> mn = np.zeros(len(nr))
>>> vr = np.zeros(len(nr))

# Generate samples and compute mean and variance
>>> for i in range(len(nr)):
>>>     n = int(nr[i])
>>>     x = powerlaw.Power_Law(xmin=1, parameters=[alpha]).generate_random(n)
>>>     mn[i] = np.mean(x)
>>>     vr[i] = np.var(x)

# Define the figure
>>> fig, ax = plt.subplots(figsize=(5, 4))

# Plot data
>>> ax.loglog(nr, mn, 'o', linewidth=2, color='peachpuff', markersize=8)
>>> ax.loglog(nr, vr, 'o', linewidth=2, color='skyblue', markersize=8)

# Set plot properties
>>> ax.set_facecolor('white')
>>> ax.set_xlim([np.min(nr), np.max(nr)])
>>> ax.set_ylim([np.min([np.min(mn), np.min(vr)]), np.max([np.max(mn), np.max(vr)])])
>>> ax.set_xticks(10.0**np.arange(1, 7), minor=False)
>>> ax.set_yticks(10.0**np.arange(0, 10), minor=False)
>>> ax.tick_params(labelsize=12)
>>> ax.spines['left'].set_color('black')
>>> ax.spines['left'].set_linewidth(2)
>>> ax.spines['bottom'].set_color('black')
>>> ax.spines['bottom'].set_linewidth(2)
>>> ax.spines['right'].set_color('none')
>>> ax.spines['top'].set_color('none')
>>> ax.xaxis.tick_bottom()
>>> ax.yaxis.tick_left()

# Set labels and title
>>> ax.set_xlabel('Size of sample, n', fontsize=10)
>>> ax.set_ylabel('Value', fontsize=10)
>>> ax.set_title('Sample Mean and Variance of Power-law Distributed Random Variables', fontsize=12)

# Add legend and text
>>> ax.legend(['Sample mean', 'Sample variance'], fontsize=8)
>>> ax.text(8*10**4, 10**8, r'$\alpha={}$'.format(alpha), fontsize=10, bbox=dict(facecolor='white', edgecolor='black', boxstyle='round,pad=0.5'))

# Show the plot
>>> plt.show()
```

Output

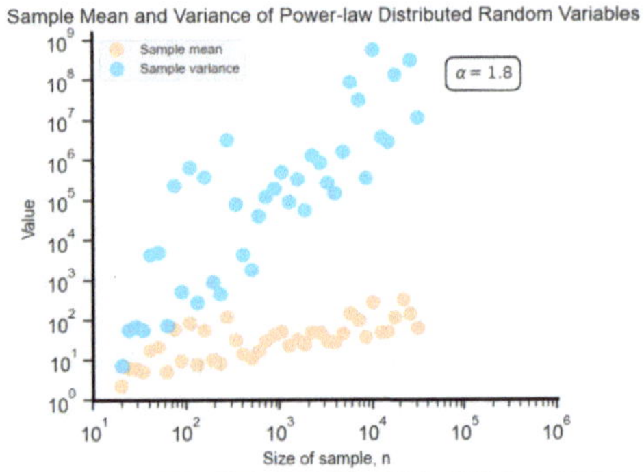

The plot in the output represents the mean values and variances of samples of random variables distributed according to a power-law. On the *x-axis*, we have the sample sizes, on a logarithmic scale. On the *y-axis*, we have the sample mean and variance, also on a logarithmic scale. Each point represents a sample. The color of the points differentiates mean values (orange) from variances (blue). The legend in the upper left corner shows which quantities correspond to each color.

4.6.5 Power-Law Estimation from Empirical Data

The initial objective is to accurately estimate empirical values within a power-law distribution. Studies on empirical distributions following a power-law typically provide estimates of the scale parameter (α) and the lower bound of the scaling region (x_{min}). By taking the logarithm on both sides of the power-law equation, we observe the distribution follows the formula:

$$\ln p(x) = -\alpha \ln x + c \tag{4.32}$$

as a straight line on a log-log plot.

According to Clauset et al. (2009), a common method to test for power-law behavior is to measure the quantity of interest (x), construct a histogram representing its frequency distribution, and plot it on a log-log scale. If the plot roughly follows a straight line, it can be inferred that the distribution follows a power-law, with a scale parameter (α) given by the absolute slope of the line. The slope is extracted by performing a least squares linear regression on the logarithmically transformed histogram.

To estimate α correctly, a value for the lower limit x_{min} of power-law behavior in the data is required. We assume this value is known for now. The method used to estimate a parameterized model like the power-law is the maximum likelihood method, which can provide accurate parameter estimates within large datasets. Assuming the data is derived from a distribution that follows an exact power-law for $x > x_{min}$, we can derive the maximum likelihood estimators (MLEs) for the scale parameter for both the discrete and continuous cases. The MLE for the continuous case is described as follows:

$$\hat{\alpha} = 1 + n\left[\sum_{i=1}^{n} \ln \frac{x_i}{x_{\min}}\right]^{-1} \tag{4.33}$$

where x_i with $i = 1, 2, \ldots, n$ represents the observed values of x where $x_i \geq x_{min}$.

Before estimating the scale parameter (α), all samples below the value of x_{min} must be removed, leaving only the values for which the power-law model is considered to be valid. To obtain an accurate estimate of α, a method for estimating an accurate value of x_{min} is required. The maximum likelihood estimation (MLE) gives an accurate value when x_{min} is chosen close to the actual value and deviates quickly below this point (because the distribution deviates from a power-law) and more slowly above it (due to a decrease in sample size).

The most common estimation method for x_{min} can be applied for both continuous and discrete data. The idea is to choose a value of x_{min} that makes the empirical distribution of the data and the "best-fitting" power-law model above x_{min} as similar as possible. The best measure to quantify the distance between two probability distributions is the Kolmogorov–Smirnov (or KS statistic), which simply measures the maximum distance between the cumulative distribution function of the data and its estimated model:

$$D = \max_{x \geq x_{\min}} \left| S(x) - P(x) \right| \tag{4.34}$$

where $S(x)$ is the cumulative distribution function of the data for the observations with values of at least x_{min}, and $P(x)$ is the cumulative density function of the power-law model that best represents the data in the region where $x \geq x_{min}$. Therefore, the estimated value x_{min} is the value of x_{min} that minimizes D.

Given a set of data and a hypothesized power-law distribution from which the data originated, the primary objective is to assess the plausibility of this hypothesis. A conventional approach involves using goodness-of-fit tests, which yield a p-value quantifying the plausibility of the hypothesis.

These tests are based on measuring the distance between the empirical data distribution and the hypothesized model. This distance is then compared against the measures of distance for comparable synthetic data from the same model, and the p-value is defined as the fraction of the synthetic distance greater than the empirical distance.

If the p-value is large (approaching 1), the difference between the empirical data and the model can be attributed to statistical fluctuations alone. Conversely, if the p-value is small, the model is unlikely to be a plausible fit for the data (Clauset et al., 2009).

In essence, goodness-of-fit tests and p-value assessments form an integral part of determining whether a given model, such as a power-law distribution, accurately represents a dataset. By comparing empirical data to synthetically generated data and evaluating statistical deviations, researchers can either validate or challenge their initial hypotheses, leading to more robust and reliable findings.

The following code aims to generate and analyze data derived from a power-law distribution. The code then fits the generated data to a power-law distribution using the *Fit* function from the *powerlaw* library in R and Python. This function estimates the minimum value (x_{min}) from which the data follows a power-law distribution and the α value for the fitted power-law distribution. After fitting the data, the code compares the fit to other types of distributions; this comparison differs in the two examples provided for R and Python. A log-likelihood ratio test is used, with the results expressed as R (the ratio of the likelihoods) and p (the significance value).

Lastly, the code plots the Probability Density Function (PDF) of the generated data along with the PDFs of the fitted power-law and the other distributions.

The most common distributions, taking into account whether the variable is continuous or discrete, are listed in Table 4.1.

Table 4.1 Main common distributions

Types of distribution	px/p(x)
Power-law (discrete)	$x^{-\alpha}/\zeta(\alpha)$
Power-law (continuous)	$\alpha x^{-\alpha}$
Poisson (discrete)	$e^{-\mu}\mu^{x}/x!$
Exponential (discrete)	$(1 - e^{-\lambda})e^{-\lambda x}$
Exponential (continuous)	$\lambda e^{-\lambda x}$
Truncated power-law (continuous)	$\frac{\lambda^{1-\alpha}}{\Gamma(1-\alpha)} x^{-\alpha} e^{-\lambda x}$
Log-normal (continuous)	$\frac{1}{x\sqrt{2\pi\sigma^2}} e^{-(\ln x-\mu)^2/(2\sigma^2)}$
Stretched exponential (continuous)	$\beta\lambda^{\beta} x^{\beta-1} e^{-(\lambda x)^{\beta}}$

4.6.6 R Code and Output

```
# Install powerlaw if not already installed
# Import the library
> library("poweRlaw")

# Set parameters
> xmin = 1 # minimum value
> alpha = 2.5 # scaling parameter
> n = 100 # number of observations

# Generate power-law distributed random numbers (continuous values)
# For discrete values: data = rpldis(n, xmin = xmin, alpha = alpha)
> data = rplcon(n, xmin = xmin, alpha = alpha)
> data<-as.data.frame(data)

# Create continuous power-law object
> pl = conpl$new(data$data)

# Estimate xmin and set for the power-law object
> est = estimate_xmin(pl)
> pl$setXmin(est)

# Plot the data
> plot(pl, pch=21, bg=2, panel.first=grid(col="grey80"), xlab="X", ylab="CDF" , cex = 0.6 )
legend(x = "topright", legend = c("PowerLaw = red", "Lognormal = green", "Exponential = blue", "Stretched exponential =
Violet"), cex=0.6, lty = c(1, 1, 1,1), col = c("red", "green", "blue", "darkorchid2"))

# Fit a line to the power-law data
> lines(pl, col=2, lwd=2)

# Repeat process for log-normal, exponential and Weibull distributions
> ln = conlnorm$new(data$data)
> est1 = estimate_xmin(ln)
> ln$setXmin(est1)
> lines(ln, col=3, lwd=2)

> exp = conexp$new(data$data)
> est2 = estimate_xmin(exp)
> exp$setXmin(est2)
> lines(exp, col=4, lwd=2)

> we = conweibull$new(data$data)
> est4 = estimate_xmin(we)
> we$setXmin(est4)
> lines(we, col=6, lwd=2)
```

Output

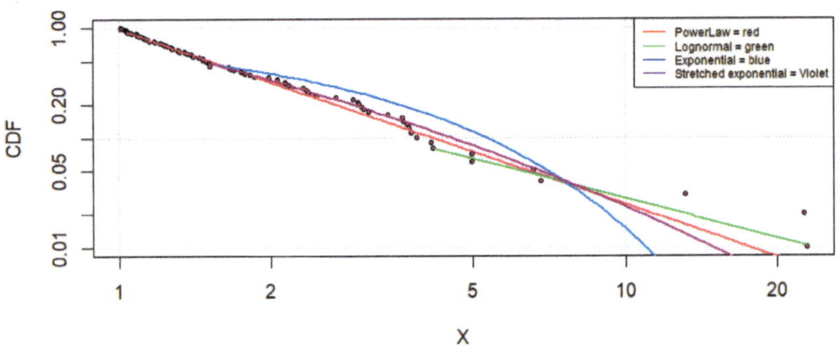

```
# Bootstrap the power-law model
> bs = bootstrap(pl, no_of_sims=1000, threads=4)
> sd(bs$bootstraps[,2])

# Bootstrap log-normal, exponential and Weibull models
> bs1 = bootstrap(ln)

> bs2 = bootstrap(exp)
> bs3 = bootstrap(we)

# Calculate bootstrap p-value
> bs_p = bootstrap_p(pl)
> bs_p$p

# Plot the p-value
> plot(bs_p, trim=0.1)

# Compare power-law distribution with log-normal, exponential and Weibull distributions
> ln$setXmin(pl$getXmin())
> est = estimate_pars(ln)
> ln$setPars(est)
> comp = compare_distributions(pl, ln)

> exp$setXmin(pl$getXmin())
> est1 = estimate_pars(exp)
> exp$setPars(est1)
> comp1 = compare_distributions(pl, exp)

> we$setXmin(pl$getXmin())
> est2 = estimate_pars(we)
> we$setPars(est2)
> comp2 = compare_distributions(pl, we)

# Estimate parameters for each distribution
> pl$setPars(estimate_pars(pl))
> ln$setPars(estimate_pars(ln))
> exp$setPars(estimate_pars(exp))
> we$setPars(estimate_pars(we))

# Compare each distribution with power-law distribution
> comp = compare_distributions(pl, ln)
> comp$p_two_sided

> comp = compare_distributions(pl, exp)
> comp$p_two_sided

> comp = compare_distributions(pl, we)
> comp$p_two_sided
```

Output

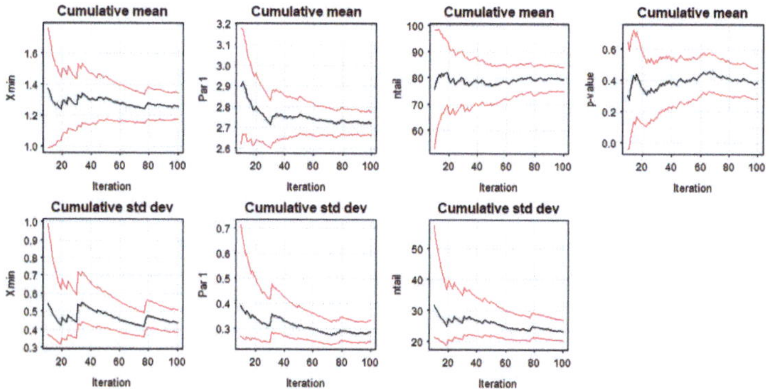

The script presents a comprehensive overview of fitting and comparing various distribution models with the synthetic data. It initiates by fitting the data to a power-law distribution model and visually presents this fit along with the synthetic data. Further, it compares the power-law fit with log-normal, exponential, and Weibull distributions, providing a holistic graphical representation of the various fits. Next, the script implements resampling statistics by implementing bootstrap procedures for all four distribution models. The bootstrap process generates multiple new datasets from the original one, enabling a reliable estimation of variability for a statistic, increasing the robustness to the analysis. The subsequent part of the script involves computing p-values, which give a statistical measure of the goodness-of-fit for the bootstrap models. It does this by comparing the original synthetic data distribution with the distribution of the bootstrapped data. A lower p-value signifies a poor fit between the data and the model, while a high p-value (closer to 1) implies that any differences can be attributed solely to statistical fluctuations, indicating a plausible fit.

The script then shifts its focus to estimating the parameters of all considered distributions and compares these with the parameters of the power-law model. The p-values derived from this comparison give a statistical measure of the similarity between the distributions.

4.6.7 Python Code and Output

```
# Install powerlaw if not already installed
>>> !pip install powerlaw

# Import necessary libraries
>>> import powerlaw
>>> import matplotlib.pyplot as plt
>>> import numpy as np

# Define power-law parameters
>>> xmin = 1.0
>>> alpha = 2.5
>>> num_samples = 10000

# Generate power-law distributed data
>>> data = powerlaw.Power_Law(xmin=xmin, parameters=[alpha]).generate_random(num_samples)

# Fit the data to a power-law distribution, allowing the fit to choose its own xmin
>>> fit = powerlaw.Fit(data)

>>> print(f"Estimated xmin: {fit.xmin:.2f}")
>>> print(f"Estimated alpha: {fit.alpha:.2f}")

# Compare the power-law fit to other distributions (lognormal_positive, exponential, stretched_exponential)
>>> R, p = fit.distribution_compare('power_law', 'lognormal_positive')
>>> print(f"\nLognormal comparison, R: {R:.2f}, p: {p:.2f}")

>>> R, p = fit.distribution_compare('power_law', 'exponential')
>>> print(f"\nExponential comparison, R: {R:.2f}, p: {p:.2f}")

>>> R, p = fit.distribution_compare('power_law', 'stretched_exponential')
>>> print(f"\nStretched exponential comparison, R: {R:.2f}, p: {p:.2f}")

# Create histogram of data
>>> counts, bins = np.histogram(data, bins=np.logspace(np.log10(min(data)), np.log10(max(data)), 50), density=True)

# Plot the data as points (circles)
>>> bin_centers = (bins[:-1] + bins[1:]) / 2
>>> plt.scatter(bin_centers, counts, color='orange', s=8)

# Plot the PDFs of the power-law, lognormal_positive, exponential, and stretched exponential fits
>>> fit.power_law.plot_pdf(color='b', linestyle='-', linewidth=1)
>>> fit.lognormal_positive.plot_pdf(color='r', linestyle='-', linewidth=1)
>>> fit.exponential.plot_pdf(color='g', linestyle='-', linewidth=1)
>>> fit.stretched_exponential.plot_pdf(color='purple', linestyle='-', linewidth=1)

>>> plt.title('Data and Fits')
>>> plt.xlabel('Data')
>>> plt.ylabel('PDF')
>>> plt.xscale('log')
>>> plt.yscale('log')
>>> plt.legend(['Data', 'Power law', 'Lognormal', 'Exponential', 'Stretched Exponential'])
>>> plt.show()
```

Output

Estimated x_{min}: 1.03
Estimated alpha: 3.01

Lognormal comparison, R: 626.70, p: 0.00

Exponential comparison, R: 1554.79, p: 0.00

Stretched exponential comparison, R: 26.47, p: 0.00

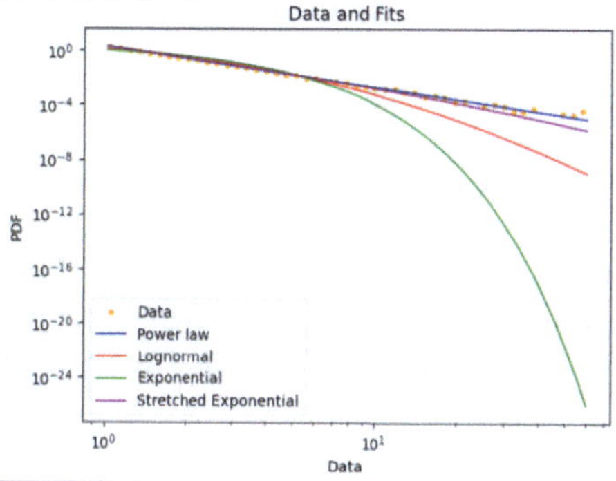

The script estimates the x_{min} value as 1.03, which means that only data points greater than or equal to this value follow the power-law distribution. The estimated α value for the power-law distribution is 3.01. In comparing the power-law fit to other distributions, the script finds that the power-law fit is significantly better than the lognormal, exponential, and stretched exponential fits. This is indicated by the large positive R values (626.70, 1554.79, and 26.47, respectively) and the p-values of 0.00.

A positive R value means that the data is more likely to follow the first distribution (in this case, the power-law distribution). A p-value of 0.00 indicates that the result is statistically significant, suggesting that the differences between the fits are not due to random chance.

To summarize, the data generated in this code is best described by a power-law distribution with $x_{min} = 1.03$ and $\alpha = 3.01$, and this distribution fits the data significantly better than the lognormal, exponential, and stretched exponential distributions.

4.7 Case Study Using Python: Exploring Citation Distribution Through the Lens of Power-Law

The count of citations garnered by a scientific work or an individual researcher has emerged as a significant metric for evaluating quality in the scientific community. While it may not provide a precise assessment of the importance of a given scientific paper or researcher, citation count can be viewed as an objective and reasonable benchmark (Reale & Zinilli, 2017). The unequal distribution of citations was presumed to be a result of the "Matthew effect" in scientific research, where papers garner additional citations partially due to their existing citations (Merton, 1968). This occurrence can be attributed to the tendency of researchers to find earlier cited papers from the bibliographies of other articles.

However, it has been observed that academic papers accumulate citations at a rate that exceeds the expectations set by the standard lifecycle of a publication. This indicates that the theory of preferential attachment may not provide a comprehensive explanation for the observed trend of citation accumulation. This is due, in part, to the fact that a minority of authors have the capacity to cite a scholarly article with full awareness of the precise number of citations that article will have received by the time their own work is published. In an attempt to further study this phenomenon, a specific study has unearthed evidence from the discipline of physics which suggests that the level of interest an article garners experiences an exponential decline over a given period (Eom & Fortunato, 2011). The distribution of citations to scientific papers is often described by a power-law (Redner, 2005; Brzezinski, 2015). This means that a small number of papers receive a lot of citations, while the majority receive very few or none at all. The broad interest in this distribution and its growth mechanisms have led to critical debates and substantial research work.

The dominant viewpoint suggests that the degree distributions in complex networks adhere to a power-law dependence, characterized by an exponent that lies within the range of 2–3 ($2 \leq \alpha \leq 3$) (Zinilli, 2016).

Recent studies have addressed the empirical detection of power-law behavior in citation distribution. These studies have provided substantial evidence on the power-law behavior of citation distributions, leading to several conclusions. Primarily, it is essential to note that the power-law behavior does not apply uniformly across all fields. (Clauset et al., 2009; Zinilli, 2016). In contrast to previous research, recent academic works have identified that the scaling factor of the power-law distribution predominantly exceeds 3.5, providing a new perspective on this topic (Brzezinski, 2015). Finally, it should be noted that power-law in citation distributions accounts for merely around 2% of the most frequently cited articles.

Hence, the primary aim of this case study, employing a statistically robust methodology, is to discern if the power-law model is the optimal descriptor for the observed distribution of cited papers. This raises the question: does the power-law model indeed provide the most accurate representation of citation distribution in cited articles?

Our objective involves conducting a comprehensive assessment of the power-law's applicability to the data in question, juxtaposing its efficacy against a range of alternative statistical distributions. It's critical to acknowledge that the types of distributions we select for comparison are contingent upon the inherent attributes of the data we are working with. Particularly, we must discern whether the data points are characterized as continuous or discrete variables. We will consider a range of candidate distributions for both data types and employ rigorous statistical methodologies to discern the most accurate representation of our data. Our ultimate goal is to identify the statistical distribution, be it the power-law or otherwise, that provides the most accurate and reliable description of our citation data. In particular, we test power-law versus lognormal, exponential, truncated power-law and stretched exponential distribution.

If citations follow a log-normal distribution, it suggests that the multiplicative effects of many independent and identically distributed random variables play a significant role in the citation process. A paper would receive a citation not solely based on its quality, but due to a combination of factors, such as visibility, the influence of the authors, or the popularity of the topic. The log-normal distribution reflects multiplicative random processes with a skewed distribution, meaning a majority of papers receive few citations, while a small fraction receive a lot. If citations follow an exponential distribution, it indicates a constant hazard rate or a memoryless process. This would mean that a paper's probability of receiving an additional citation does not depend on the number of citations it already has, which contradicts the concept of the "rich-get-richer" dynamic, or preferential attachment, often seen in citation behavior. A truncated power-law is a power-law distribution that has an upper limit or cutoff beyond which no values can occur. In citation data, a truncated power-law suggests that there is a maximum number of citations that any given paper can receive, likely due to constraints such as the total number of papers being published or the focus of current research trends. On the other hand, a stretched exponential distribution (also known as the Weibull distribution) is a generalization of the exponential distribution that allows for a varying rate of decay. In the context of citation data, a stretched exponential distribution may suggest that the rate at which papers accumulate additional citations changes over time. This could be due to a variety of factors, such as a paper's topic becoming more or less popular, the overall growth of the scientific community, or changes in how researchers find and cite earlier work (Zinilli, 2016). All these distributions tested provide different insights into the mechanisms governing citations, and finding the best fit for our data can help us understand the underlying dynamics more accurately.

The citation count in question refers to the total number of times the work of the 1120 most frequently cited physicists between 1998 and 2014 was referenced in other academic publications. This dataset was sourced from Scopus, one of the largest abstract and citation databases of peer-reviewed literature.

The dataset provides an insight into the impact and relevance of these physicists' contributions during this 16-year period. It helps evaluate the influence of these individuals within their field, as frequent citation often implies that the work has been influential or foundational within its discipline.

Data and Methodology

The citation count refers to the enumeration of instances of research conducted by the top 1120 physicists from 1998 to 2014. This data ranges from a high of 16,994 to a low of 2328 citations.

The spread of the citation count demonstrates the variance in the impact of the works produced by these physicists over the 16-year period. An intriguing aspect that can be gathered from this data is the distribution of citation counts, which can be understood using a Kernel Density Estimation (KDE) (Fig. 4.1). The Kernel is a non-parametric method of estimating the probability density function of a random variable, in this case, citation counts. By applying KDE, we can better visualize and understand the distribution of citation counts, allowing us to identify patterns or trends, such as which citation counts are most common among these top-cited physicists. This can provide valuable insights into the citation patterns and the influence of these physicists' work within the scientific community.

The wide range of citation counts suggests there is substantial variability in the data. The dataset follows a skewed distribution, common in citation data, where a small number of papers or researchers receive a large number of citations, while the majority receive fewer citations. This could be influenced by a range of factors such as the specific subfield of physics, the novelty or significance of the research, and the extent of collaboration with other researchers. Furthermore, the very high counts for some physicists suggest that their work has had a substantial impact on their field, influencing a large number of subsequent research studies. These individuals may have made particularly ground-breaking or significant contributions to their field, causing their work to be frequently cited.

In the next paragraph, titled "Results", the process implemented by Clauset et al. (2009) is presented, which offers a good method to analyze and understand the

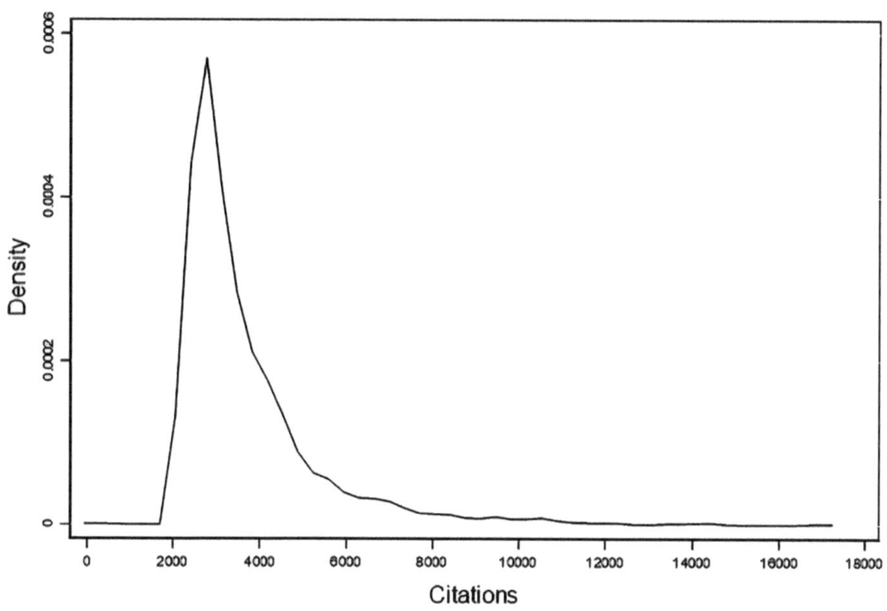

Fig. 4.1 Kernel Density Estimation for citation data

distribution of citations. This framework can be used to verify whether the number of citations received by scientific papers follows a power-law distribution. A power-law distribution would suggest that a small number of papers receive a large majority of citations, while the vast majority of papers are less frequently cited.

The methodology of Clauset et al. (2009) goes further by allowing us to compare the fit of the power-law model against various alternative models. This is accomplished using a statistical method known as a likelihood ratio test, which can inform us whether one model is statistically better than another at describing the data. In applying this method to citation data, we could ascertain if a power-law distribution is indeed the most appropriate model for this data. This, in turn, could provide profound insights into the dynamics of scientific recognition and inform strategies aimed at encouraging a more balanced distribution of citations across scientific research.

Results

In the forthcoming section, the intention is to test the model proposed by Clauset et al. (2009) using our citation data. We shall demonstrate each step accurately, leveraging Python as our primary computational tool. All steps, from data preprocessing to the fitting and evaluation of the model, will be thoroughly explained and demonstrated, providing an in-depth insight into the application of the method.

Input

```
# Import necessary libraries
>>> from __future__ import division
>>> import powerlaw
>>> import numpy as np
>>> from mpl_toolkits.mplot3d import Axes3D
>>> import matplotlib.pyplot as plt

# Set numpy to ignore division errors and invalid value operations
>>> np.seterr(divide='ignore', invalid='ignore')

# Load the data from the "Citations_data" file and convert it to a numpy array
# Replace 'Data.get("Citations_data")' with the actual function that reads your data
>>> data = np.loadtxt("path\\Citations_data.txt")

# Fit the power-law distribution to the data using the maximum likelihood method
>>> results = powerlaw.Fit(data, fit_method='Likelihood', discrete=True, estimate_discrete=
True, discrete_approximation='round')

# Display the parameters of the fitted power-law
>>> print(results.xmin)          # minimum data value for the power-law fit
>>> print(results.fixed_xmin) # if the xmin value was held fixed
>>> print(results.power_law.alpha) # the scaling exponent (alpha) of the power-law fit
>>> print(results.power_law.D) # the Kolmogorov-Smirnov distance between the data and the
fit
```

Output

```
# The scaling exponent (alpha) of the power-law fit
4.08

# The Kolmogorov-Smirnov distance between the data and the fit
0.03
```

The above measures are the α (scaling parameter) and the D (Kolmogorov-Smirnov distance) of the power-law fit. The scaling exponent (α) in a power-law distribution describes the rate at which the probability of observing a value decreases as the value increases. Larger alpha values generally imply a faster decay in the probability of high values.

The Kolmogorov-Smirnov (KS) distance is a measure of the goodness-of-fit. It quantifies the maximum difference between the cumulative distribution functions of the empirical data and the fitted power-law distribution. Smaller values of D suggest a better fit to the power-law distribution.

Input

```
# Repeat the fitting process and print the parameters.
>>> results = powerlaw.Fit(data, fit_method='Likelihood', discrete=True,
estimate_discrete=True, discrete_approximation='round')
>>> print(results.xmin)
>>> print(results.fixed_xmin)
>>> print(results.power_law.alpha)
>>> print(results.power_law.D)

# Fit power-law to the data without specifying the fit method. This time the results are stored in the
variable 'fit'.
>>> fit = powerlaw.Fit(data, fit_method='Likelihood', discrete=True, estimate_discrete=True,
discrete_approximation='round')

# Display the parameters of the fitted power-law
>>> print(fit.xmin)
>>> print(fit.fixed_xmin)

# Print the names of the first parameters of different fitted distributions
>>> print(fit.power_law.parameter1_name)
>>> print(fit.lognormal.parameter1_name)
>>> print(fit.exponential.parameter1_name)

# Calculate and print the log-likelihood ratio and p-value between power-law and other distributions
>>> R, p = fit.loglikelihood_ratio('power_law', 'lognormal', normalized_ratio=True)
>>> print(R, p)
>>> R, p = fit.loglikelihood_ratio('power_law', 'exponential', normalized_ratio=True)
>>> print(R, p)
>>> R, p = fit.loglikelihood_ratio('power_law', 'truncated_power_law', nested=True)
>>> print(R, p)
>>> R, p = fit.loglikelihood_ratio('power_law', 'stretched_exponential', nested=True)
>>> print(R, p)
```

Output

```
R= -1.32 p=0.18
R=0.72 p=0.47

# Assuming nested distributions
R=-2.55 p=0.02
R=-2.23 p=0.035
```

The *loglikelihood_ratio* function in the *powerlaw* package is used to compare the goodness-of-fit between two distribution models. This is done by calculating the log-likelihood ratio R and the p-value p. When comparing the power-law and log-normal distributions, the log-likelihood ratio was -1.32 and the p-value was 0.18. The negative log-likelihood ratio suggests that the lognormal model provides a better fit to the data compared to the power-law model. However, the p-value exceeding 0.05 indicates that this difference in fit is not statistically significant at the 95% confidence level.

For the power-law and exponential distributions, the log-likelihood ratio was 0.72 and the p-value was 0.47. The positive ratio implies that power-law model fits the data more adequately than the exponential model. Yet, similar to the previous comparison, the p-value is above 0.05, suggesting that the difference in the fit of these two models to the data is not statistically significant at the 95% confidence level.

Comparing the power-law and truncated power-law distributions yielded a log-likelihood ratio of −2.55 and a p-value of 0.02. In this case, the negative ratio indicates a better fit for the truncated power-law model. Unlike the previous two comparisons, the p-value is below 0.05, which demonstrates a statistically significant difference in the fit of the power-law and truncated power-law models to the data at the 95% confidence level.

Finally, the comparison between the power-law and stretched exponential distributions resulted in a log-likelihood ratio of −2.23 and a p-value of 0.035. Here again, the negative ratio suggests that the stretched exponential model fits the data more accurately than the power-law model. As the p-value is below 0.05, this difference is statistically significant at the 95% confidence level.

In summary, while a positive log-likelihood ratio indicates a superior fit for the power-law model, a negative value implies the secondary model is a more suitable fit. The p-value is a statistical measure of the confidence in this difference in fit: a p-value less than 0.05 typically indicates a statistically significant difference.

Input

```
# Plot the complementary cumulative distribution functions (CCDF) of empirical data and fitted
models
>>> fig1 = fit.plot_ccdf(linewidth=3, label='Empirical Data')
>>> fit.power_law.plot_ccdf(ax=fig1, color='black', linestyle='-.',label='Power Law',
linewidth=0.8)
>>> fit.lognormal.plot_ccdf(ax=fig1, color='g', linestyle='--',label='Lognormal')
>>> fit.exponential.plot_ccdf(ax=fig1, color='y', linestyle='--', label= 'Exponential', linewidth=
0.8)
>>> fit.truncated_power_law.plot_ccdf(ax=fig1, color='m', linestyle=':', survival=True,
label='Power Law + cut-off', linewidth=2)
>>> fit.stretched_exponential.plot_ccdf(ax=fig1, color='b', linestyle='--',label=
'Stretched_exponential', linewidth = 0.8)

# Set the labels for the plot axes and the legend
>>> fig1.set_ylabel(r"$p(X\geq x)$")
>>> fig1.set_xlabel(r"Degree")
>>> plt.xlabel("Citations Distribution --> Blue = Empirical Data")
>>> handles, labels = fig1.get_legend_handles_labels()
>>> fig1.legend(handles, labels, loc=3, fontsize=9)

# Set the title of the plot and save the figure
>>> plt.title("Comparison of Power Law with Other Fits")
>>> plt.savefig("powerlaw.png")
```

Output

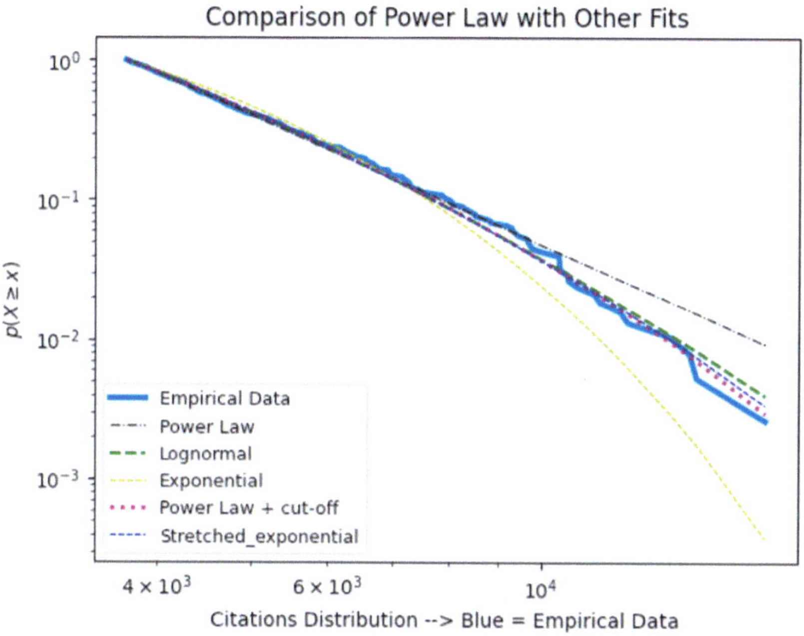

The generated plot, the complementary cumulative distribution functions (CCDFs) of the empirical citation data and several theoretical distribution models are visually compared. The aim is to illustrate which model most closely captures the actual distribution of citations across papers.

The empirical data, depicted by a solid blue line, forms the benchmark against which all models are evaluated. This line traces the observed distribution of citations, presenting a "ground truth" scenario.

The dashed black line represents the power-law model. Visual inspection allows for an assessment of its fit to the empirical data. Notably, if the power-law model were a perfect fit, this line would overlap with the blue line of the empirical data. The lognormal model is denoted by a green dashed line. By comparing its trajectory with the blue empirical data line, one can gain a sense of how this model captures the citation distribution. The closer the lines, the better the fit. Similarly, the yellow dashed line represents the exponential model. The divergence of this line from the empirical data clearly indicates that this distribution is less suited to accurately represent the citation dynamics.

The truncated power-law model, represented by a magenta dotted line, introduces a new element into the mix—the possibility of a saturation point in citation counts. The proximity of this line to the empirical data, compared to the original power-law line, implies that the truncated model provides a more accurate representation of the citations. Finally, the stretched exponential model is depicted by a dashed blue line. This model, which allows for a rate of decay that changes over time, is another candidate for representing the citation dynamics. By visually comparing its proximity to the empirical data line with that of the other model lines, we can gain a sense of how well the stretched exponential model fits the data. While the plot offers an intuitive method for comparing these models, it is essential to remember that it only provides a qualitative perspective. For a rigorous and quantitative evaluation of model fit, the application of statistical comparison methods, such as log-likelihood ratios, is fundamental.

4.8 Conclusions

In this case study, we applied different probability distributions to the citation dataset. Each citation represents a reference to a paper in another paper. Analysis shows that the distribution of scientific citations is a complex and multifaceted phenomenon, that cannot be fully captured by a single statistical model.

The initial exploration using the power-law model suggested a "rich get richer" scenario where a small number of papers accumulate a large number of citations. However, further investigation with lognormal and exponential models revealed a more nuanced landscape. Although these models did not significantly outperform the power-law model, they suggested alternative narratives for the distribution of citations among papers. More compelling insights emerged when we compared the power-law model to the truncated power-law and stretched exponential models. Both models provided a significantly better fit, indicating they may more accurately represent the dynamics of the citation landscape. The truncated power-law model suggested a saturation effect in citations, where there is an upper limit to the number of citations a paper can receive (even the most influential papers cannot garner more than a certain number of citations), while the stretched exponential model indicated a rapid decay in the likelihood of a paper achieving high citation counts (very high citation counts are even less likely than one might expect). These findings illuminate the diverse and complex nature of the citation landscape in scientific literature. They emphasize the need for a nuanced approach when examining citation dynamics and caution against oversimplified interpretations.

References

Ahn, Y. Y., Bagrow, J., & Lehmann, S. (2010). Link communities reveal multiscale complexity in networks. *Nature, 466*, 761–764. https://doi.org/10.1038/nature09182

Albert, R., & Barabási, A.-L. (2000). Topology of evolving networks: Local events and universality. *Physical Review Letters, 85*, 5234–5237.

Albert, R., & Barabási, A. L. (2002). Statistical mechanics of complex networks. *Reviews of Modern Physics, 74*(1), 47.

Barabási, A. L., & Albert, R. (1999). Emergence of scaling in random networks. *Science, 286*(5439), 509–512.

Barabási, A.-L., Jeong, H., Neda, Z., Ravasz, E., Schubert, A., & Vicsek, T. (2002). Evolution of the social network of scientific collaborations. *Physica A: Statistical Mechanics and Its Applications, 311*, 590–614.

Barrat, A., & Weigt, M. (2000). On the properties of small-world network models. *The European Physical Journal B-Condensed Matter and Complex Systems, 13*, 547–560.

Battiston, F., Cencetti, G., Iacopini, I., Latora, V., Lucas, M., Patania, A., & Petri, G. (2020). Networks beyond pairwise interactions: Structure and dynamics. *Physics Reports, 874*, 1–92.

Blondel, V. D., Guillaume, J. L., Lambiotte, R., & Lefebvre, E. (2008). Fast unfolding of communities in large networks. *Journal of Statistical Mechanics: Theory and Experiment, 2008*(10), P10008.

Borgatti, S. P., & Everett, M. G. (2000). Models of core/periphery structures. *Social Networks, 21*(4), 375–395.

Brzezinski, M. (2015). Power laws in citation distributions: Evidence from Scopus. *Scientometrics, 103*, 213–228. https://doi.org/10.1007/s11192-014-1524-z

Clauset, A., Shalizi, C. R., & Newman, M. E. (2009). Power-law distributions in empirical data. *SIAM Review, 51*(4), 661–703.

Cordoba, J. (2008). On the distribution of city sizes. *Journal of Urban Economics, 63*, 177–197.

Dorogovtsev, S. N., & Mendes, J. F. F. (2001). Scaling properties of scale-free evolving networks: Continuous approach. *Physical Review E, 63*, 056125.

Dorogovtsev, S. N., Mendes, J. F. F., & Samukhin, A. N. (2000). Structure of growing networks: Exact solution of the Barabási–Albert model. *Physical Review Letters, 85*, 4633–4636.

Eom Y.-H., & Fortunato S. (2011). Characterizing and Modeling Citation Dynamics. PLoS ONE 6(9):e24926. https://doi.org/10.1371/journal.pone.0024926.

Erdős, P., & Rényi, A. (1960). On the evolution of random graphs. *Publication of the Mathematical Institute of the Hungarian Academy of Sciences, 5*(1), 17–60.

Fortunato, S. (2010). Community detection in graphs. *Physics Reports, 486*(3–5), 75–174.

Fortunato, S., & Hric, D. (2016). Community detection in networks: A user guide. *Physics Reports, 659*, 1–44.

Fortunato, S., & Newman, M. E. (2022). 20 years of network community detection. *Nature Physics, 18*(8), 848–850.

Gabaix, X. (2009). Power laws in economics and finance. *Annual Review of Economics, 1*(1), 255–294.

Girvan, M., & Newman, M. E. (2002). Community structure in social and biological networks. *Proceedings of the National Academy of Sciences, 99*(12), 7821–7826.

Jeong, H., Néda, Z., & Barabási, A.-L. (2003). Measuring preferential attachment in evolving networks. *Europhysics Letters, 61*, 567–572.

Krugman, P. (1996). *The self organizing economy*. Wiley.

Lin, Z., Yin, Y., Liu, L., & Wang, D. (2023). SciSciNet: A large-scale open data lake for the science of science research. *Scientific Data, 10*(1), 315.

Merton, R. K. (1968). The Matthew effect in science: The reward and communication systems of science are considered. *Science, 159*(3810), 56–63.

Milgram, S. (1967a). Six degrees of separation. *Psychology Today, 2*, 60–64.

Milgram, S. (1967b). The small world problem. *Psychology Today, 2*(1), 60–67.

Newman, M. E. (2003). The structure and function of complex networks. *SIAM Review, 45*(2), 167–256.

Newman, M. E. J. (2004). Fast algorithm for detecting community structure in networks. *Physical Review E, 69*, 066133.

Newman, M. E. (2005). Power laws, Pareto distributions and Zipf's law. *Contemporary Physics, 46*(5), 323–351.

Newman, M. E., Moore, C., & Watts, D. J. (2000). Mean-field solution of the small-world network model. *Physical Review Letters, 84*(14), 3201.

Reale, E., & Zinilli, A. (2017). Evaluation for the allocation of university research project funding: Can rules improve the peer review? *Research Evaluation, 26*(3), 190–198.

Redner, S. (2005). Citation statistics from 110 years of physical review. *Physics Today, 58*, 49–54.

Rosvall, M., & Bergstrom, C. T. (2008). Maps of random walks on complex networks reveal community structure. *Proceedings of the National Academy of Sciences of the United States of America, 105*, 1118–1123.

Traag, V. A., Waltman, L., & van Eck, N. J. (2019). From Louvain to Leiden: Guaranteeing well-connected communities. *Scientific Reports, 9*, 5233. https://doi.org/10.1038/s41598-019-41695-z

Travers, J., & Milgram, S. (1977). An experimental study of the small world problem. In *Social networks* (pp. 179–197). Academic Press.

Watts, D. J., & Strogatz, S. H. (1998). Collective dynamics of 'small-world'networks. *Nature, 393*(6684), 440–442.

Zinilli, A. (2016). Competitive project funding and dynamic complex networks: Evidence from projects of National Interest (PRIN). *Scientometrics, 108*(2), 633–652.

Chapter 5
Statistical Social Network Models

5.1 Introduction

The use of Network Science transcends disciplinary boundaries, presenting a multitude of challenges encountered across various disciplines. Given the wide-ranging applications of networks, numerous academic communities have contributed to the advancement of network models. Recent years have witnessed the emergence of several types of models, with statistical mechanics and computer science serving as fundamental pillars for a family of models.

As discussed in the preceding chapter, these models aim to explain the underlying mechanisms that generate networks by employing probability distributions. Examples of such models include the Barabási–Albert (BA) model proposed by Barabási and Albert (1999), and the small-world theory presented by Watts and Strogatz (1998). However, it is important to note that these models often fall short in capturing the multi-mechanistic nature of real-world social networks, as they typically focus on only a limited number of specific mechanisms. For a more comprehensive understanding of these models, consulting the work of Watts (2004) is recommended.

Another category of network models has emerged from the interests of statisticians and sociologists, and it is within this category that we will focus our discussion in this chapter. These models find primary applications in the fields of social and economic sciences, facilitating statistical inference. Statistical inference involves understanding the generative processes behind networks by considering various mechanisms that are both endogenous and exogenous to the network, rather than solely focusing on studies of probability distribution. Statistical social network models, which we will explore in this chapter, aim to describe the presence of edges in real social situations or the evolution of these edges over time. These models are often applied to datasets of limited size due to significant computational complexities, which can become prohibitive with large amounts of data.

A. Zinilli, *Elements of Network Science*, Statistics and Computing,
https://doi.org/10.1007/978-3-031-84712-7_5

Key references in this area include the works of Lusher et al. (2013) and Lazega and Snijders (2015). In these models, the network itself represents the dependent variable, while the independent variables consist of node-level and dyad-level variables, as well as network configurations. Our discussion will primarily revolve around two main classes of statistical social network models: Exponential Random Graph Models (ERGMs) and Stochastic Actor-Oriented Models (SAOMs). To begin our exploration, we will examine two "traditional" statistical models that can be employed to analyze relationships between nodes in specific scenarios: the logistic model and the Quadratic Assignment Procedure (QAP). These models lay the groundwork for understanding the statistical aspects of network analysis. Subsequently, we will delve into and discuss more advanced statistical social network models, such as ERGMs and SAOMs. These models enable us to infer micro-level link formation while simultaneously describing the overall structure of the network. One notable advantage that ERGMs and SAOMs have over logistic regression models and QAP is their ability to capture network structures, such as transitivity (i.e., the tendency for friends of friends to be friends), and clustering (i.e., the tendency for nodes to form tightly-knit groups). Logistic regression struggles to effectively model these structures due to the involvement of higher-order interactions between nodes, which are not accommodated within the traditional logistic regression framework.

In conclusion, the use of Network Science is vast and encompasses multiple disciplines. The development of network models draws upon the contributions of various academic communities. While some models focus on probability distributions to explain network generation mechanisms, other models, particularly in the realm of statistical social network analysis, consider a broader range of mechanisms and emphasize statistical inference. By exploring models such as ERGMs and SAOMs, we gain insights into the formation of individual network links and the overall structure of social networks, surpassing the limitations of traditional logistic regression models and QAP.

5.2 Logistic Model

The logistic model is a type of statistical model that is used to analyze the relationship between a binary dependent variable and one or more independent variables (Robins et al., 1999). Models of logistic regression include ordered logistic regression, multinomial logistic regression, and binary logistic regression. When the dependent variable is dichotomous and the independent factors are either continuous or categorical, binary logistic regression is frequently used. A multinomial logistic regression can be used when the dependent variable has more than two groups. Instead, the ordered logit model is a regression model for an ordinal response variable. Here, we focus on models where the response variable is dichotomous. In this case, the dependent variable is usually coded as 0 or 1, where 0 represents the absence of an event (or the failure outcome) and 1 represents the presence of an event (or the success outcome). Using network data, the response variable is typically binary and represents the presence or absence of edges. The basic idea behind the logistic model is to model the probability of the binary outcome as a function of the independent variables using a logistic function. The logistic function is an S-shaped curve that has a range between 0 and 1 and is defined by the formula:

$$P(Y=1)=\frac{1}{\left(1+\exp(-z)\right)} \tag{5.1}$$

where *P(Y = 1)* is the probability of the dependent variable being equal to 1, *exp* is the exponential function, and z is the linear combination of the independent variables:

$$z=\beta_0+\beta_1 \mathrm{X}_1+\beta_2 \mathrm{X}_2+\ldots+\beta_\mathrm{k} \mathrm{X}_\mathrm{k} \tag{5.2}$$

where β_0 is the intercept term, β_1 to β_k are the coefficients for the independent variables X_1 to X_k, respectively. As a result we have:

$$P(Y=1|X)=\frac{1}{\left(1+e^{-(\beta_0+\beta_1 X_1+\beta_2 X_2+\cdots+\beta_\mathrm{k} X_\mathrm{k})}\right)} \tag{5.3}$$

The logit function is defined as the logarithm of the odds of $Y = 1$ and is given by the following equation:

$$\begin{aligned}\operatorname{logit}\left(P(Y=1|X)\right)&=\log\left(\frac{P(Y=1|X)}{1-P(Y=1|X)}\right)\\&=\beta_0+\beta_1 X_1+\beta_2 X_2+\cdots+\beta_k X_k\end{aligned} \tag{5.4}$$

where $logit(P(Y = 1 \mid X))$ is the natural logarithm of the odds of $Y = 1$ given the values of the predictor variables X. The odds of $Y = 1$ is defined as the ratio of the probability of $Y = 1$ to the probability of $Y = 0$ (i.e., the failure outcome). Thus, $odds(Y = 1 \mid X) = P(Y = 1 \mid X)/(1 - P(Y = 1 \mid X))$.

β_0, β_1, β_2, …, βk are the regression coefficients (i.e., weights) that quantify the effect of each predictor variable on the log-odds of the success outcome. X_1, X_2, …, X_k are the values of the predictor variables. The previous Eq. (5.4) is now written as:

$$\log\left(\frac{P(Y=1)}{1-P(Y=1)}\right)=\beta^T X \tag{5.5}$$

where the left hand is the log odds or logit of *P(Y)* and the right hand is the vector of predictors. Now, by raising X_i by one unit while holding the other predictors constant, the log odds change by β_j.

The logistic model estimates the values of the coefficients that maximize the likelihood of the observed data given the model. As result, maximum likelihood estimation (MLE) is commonly used to estimate the coefficients.

Let's use the simple logit model to illustrate how the likelihood of buying a product depends on its price:

$$\text{logit}(p) = \beta_0 + \beta_1 \cdot \text{price} \tag{5.6}$$

where p is the probability of purchasing the product and price is the price of the product. The logit model is expressed in terms of a logistic function, which transforms the product price into a probability of purchase. Let's say we want to estimate the coefficients of the logit model using the maximum likelihood technique and we have the price and purchase data of 10 observations. To do this, we need to calculate the likelihood of the data for each value of β_0 and β_1 and identify the values of β_0 and β_1 that maximize the likelihood of the data. The probability function of the data is first calculated. The data are distributed according to a Bernoulli distribution because the dependent variable is binary, and the likelihood function is

$$L(\beta_0,\beta_1) = \Pi\left[p^{y_i}(1-p)^{(1-y_i)}\right] \tag{5.7}$$

where p is the probability of buying the product as determined by the model, and y_i is the purchase of the product (1 if purchased is made, 0 if not). The product is calculated over all observations.

To simplify calculations, log-likelihood is used, and it is defined as:

$$\log\ L(\beta_0,\beta_1) = \Sigma\left[y_i \cdot \log(p) + (1-y_i)\cdot \log(1-p)\right] \tag{5.8}$$

with Σ equal to the sum of terms for all observations. The log-likelihood can be maximized using an optimization algorithm, such as the Newton–Raphson algorithm. The Newton–Raphson algorithm starts from an initial value of the coefficients and moves in the direction of increasing p until it finds the value of β_0 and β_1 that maximizes the log-likelihood. The Newton–Raphson algorithm is an iterative algorithm used to find the value of a root of a function. In the case of estimating the coefficients of logit models, the Newton–Raphson algorithm is used to maximize the log-likelihood function of the data. The Newton–Raphson algorithm is based on the principle that the root can be found by using knowledge of the function's slope. The algorithm specifically finds an approximation of the root using information about the slope of the function and then iterates using that estimate until the desired precision is attained. Below are the steps of the Newton–Raphson algorithm for maximizing the log-likelihood function in logit models:

1- Initialize the values of coefficients β_0 and β_1 to an arbitrary initial value.
2- Calculate the gradient vector of the log-likelihood function with respect to the coefficients:

$$\frac{\partial\left(\log L\left(\beta_0,\beta_1\right)\right)}{\partial\beta_0} = \Sigma\left[y_i - p\right] \quad (5.9)$$

$$\frac{\partial\left(\log L\left(\beta_0,\beta_1\right)\right)}{\partial\beta_1} = \Sigma\left[y_i - p\right]\cdot x_i \quad (5.10)$$

where y_i is the dependent variable (1 if a purchase is made, 0 if not), p is the probability of purchase calculated by the model, and x_i is the value of a generic independent variable associated with the dependent variable y_i.

3- Calculate the matrix of second derivatives (the Hessian matrix) of the log-likelihood function with respect to the coefficients:

$$H = \Sigma\left[p\cdot(1-p)\cdot xi^2\right] \quad (5.11)$$

where x_i is the value of the independent variable associated with the dependent variable y_i.

4- Use Newton's method (also known as the Newton–Raphson method), i.e., the step that maximizes the log-likelihood function:

$$\Delta\beta = H^{-1}\cdot\Delta\left(\log\ L\left(\beta_0,\beta_1\right)\right) \quad (5.12)$$

where $\Delta\beta$ is the Newton's method step, $\Delta(log\ L(\beta_0, \beta_1))$ is the gradient vector calculated in step 2, and H^{-1} is the inverse of the Hessian matrix calculated in step 3.

5- Update the values of the coefficients:

$$\beta_0 = \beta_0 + \Delta\beta\left[0\right] \quad (5.13)$$

$$\beta_1 = \beta_1 + \Delta\beta\left[1\right] \quad (5.14)$$

6- Up until the log-likelihood values calculated in step k and step $k-1$ differ by less than a specified tolerance threshold, repeat steps 2–5.

The Newton–Raphson algorithm is applied iteratively until the estimated coefficients reach a stable and accurate values. It is crucial to use an appropriate initial values for the coefficients as this can affect how quickly the algorithm converges. Convergence means that the coefficients of the model no longer change significantly with each iteration of the optimization algorithm. Specifically, it means that the algorithm has found the values of the coefficients that maximize the likelihood of the observed data given the model. Once the estimated coefficients have been obtained, they can be used to evaluate the effect of price on the probability of purchasing the product. For example, if the estimated price coefficient is −0.1, it means that a 1-unit increase in the price of the product reduces the probability of purchasing the product by 10%.

The logistic model can also be extended to network data. This same model just described can be used to model the probability of a connection between two nodes in a network. In this scenario, the binary response variable of a logistic model reflects the presence or absence of a link between two nodes in the network. Predictive factors can be variables related to node characteristics (node level) or variables related to the characteristics of relationships between nodes (edge level). However, when we use endogenous network variables, such as the degree centrality of a node in a network, the logit model is not appropriate because it does not take into account the interactions between nodes in the network. In econometrics, endogeneity refers to the situation in which an explanatory variable is correlated with the error term. This implies that the explanatory variable is not exogenous, meaning that it is affected not only by factors external to the model, but also by factors internal to the model. This can result in a skewed estimation of the regression parameters and a reduction in model effectiveness. The endogenous term, in a network context, as in our case, refers to the existence of reciprocal or interdependent connections between nodes. This implies that the connections between nodes are affected by the connections between other nodes in the network. Because the logit model implies that observations are independent of each other, and when examining network data, observations are interconnected and the characteristics of one node may rely on the properties of other nodes in the network, the logit model is inappropriate. For example, the degree centrality of a node depends on the number of nodes it is connected with and the degree centrality of the other nodes in the network. This means that the endogenous network variables are correlated with each other and are not independent. Furthermore, the logit model does not take into account the structure of the network, i.e., how the nodes are connected with each other. This can lead to inaccurate estimates of the coefficients of the independent variables, as the logit model assumes that all observations are equally independent and does not take into account the dependencies between them. Alternatively, network-specific regression models such as exponential random graph models can be used to analyze the relationships between endogenous network variables and dependent variable (the network). These models take into account the interactions between nodes in the network and provide more accurate estimates of the coefficients of the independent variables.

5.2.1 Stata Code and Output

// Enter the command logit followed by the dependent variable "foreign" (indicating whether the car was produced in a foreign country) and the independent variables "price" (car price) and "mpg" (miles per gallon of fuel). For example:

```
. sysuse auto.dta
. logit foreign price mpg
```

// Calculate the partial derivatives of the regression coefficients with respect to the values of the independent variables. This can be done using the "margins" command. For example, to calculate the partial derivative of the regression coefficient of "price" with respect to car price, type in:

```
. margins, dydx(price)
```

// Create a graph showing the relationship between the probability of producing a foreign car and the price of the car, keeping the variable 'mpg' constant at a specific value. This can be done using the command "marginsplot". For example, to create a graph showing the relationship between the probability of producing a foreign car and the price of the car, type:

```
. marginsplot
```

5.2.2 R Code and Output

```
# Loading the 'mtcars' dataset
> library(car)
> data(mtcars)

# Performing logistic regression. In this example, we used the glm() function to perform a logistic regression in which the dependent variable am (Automatic vs. Manual transmission) was modelled as a function of the independent variables wt (Weight in tonnes) and hp (Gross Power in horsepower). The parameter family = binomial() specifies that we are using a binomial distribution for the dependent variable, which is appropriate for logistic models.
> model <- glm(am ~ wt + hp, data = mtcars, family = binomial())

# Displaying results
> summary(model)

# Calculates odds ratios for hp
> odds_ratio <- exp(coef(model)[3])
> odds_ratio

# Define the third coefficient b
> b <- coef(model)[3]

# Calculates the partial derivative of the logit model with respect to the hp variable
> partial_derivative <- function(hp) {
  return (odds_ratio * exp(b * hp) / ((1 + odds_ratio * (exp(b * hp) - 1))^2) * b)
}

# Following steps allow to build a graph of partial derivatives
> hp_seq <- seq(from = 100, to = 150, by = 5)
> partial_derivatives <- sapply(hp_seq, partial_derivative)
> plot(hp_seq, partial_derivatives, type = "l", xlab = "hp", ylab = "Partial derivative", col = "blue")
par(font.main = 1)
title("Partial Derivative of the Logit Model", line = 0.5)
```

Output

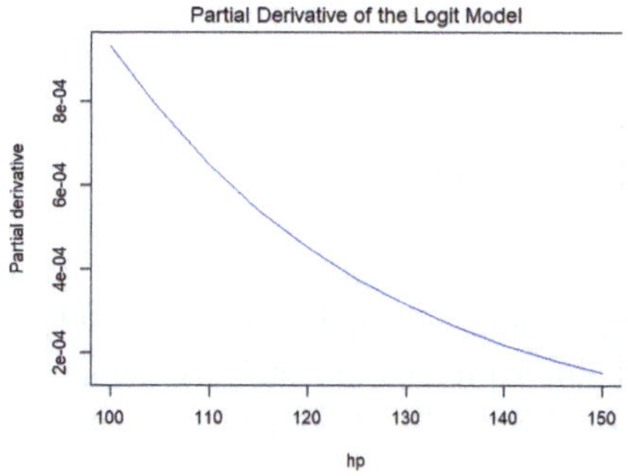

5.2.3 Python Code and Output

```
# Importing the required packages
>>> import numpy as np
>>> import pandas as pd
>>> import statsmodels.api as sm
>>> import matplotlib.pyplot as plt

# Building the dataframe with data
>>> data = pd.DataFrame({'age': [18, 21, 25, 30, 35, 40, 45, 50, 55, 60], 'prob_of_purchase': [0.05, 0.08, 0.12, 0.28, 0.35, 0.45, 
0.60, 0.75, 0.85, 0.95]})

# Definition of independent and dependent variables
>>> X = sm.add_constant(data['age'])
>>> y = data['prob_of_purchase']

# Logit model setting
>>> model = sm.Logit(y, X)

# Model Adaptation
>>> result = model.fit()

# Print results
>>> print(result.summary())

# Calculation of partial derivatives
>>> Partial_derivatives= result.params[1] * np.exp(result.params[0] + result.params[1] * data['age']) / (1 + 
np.exp(result.params[0] + result.params[1] * data['age'])) ** 2

# Plot of partial derivatives
>>> plt.plot(data['age'], Partial_derivatives)
>>> plt.xlabel('Age', fontsize=14)
>>> plt.ylabel('Partial derivative', fontsize=14)
>>> plt.title('Partial Derivatives of the Logit Model', fontsize=18)
>>> plt.grid(False)
>>> plt.show()
```

Output

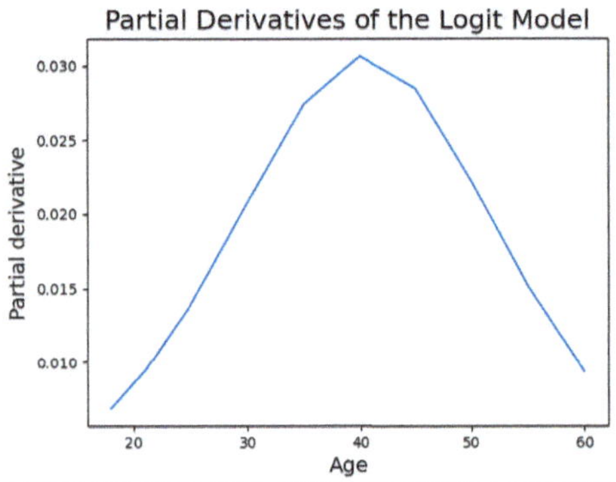

5.3 Multiple Regression QAP

Multiple Regression Quadratic Assignment Procedure (MR-QAP) is a statistical technique used to assess the impact of independent variables on a dependent variable in a network. In traditional regression analysis, the unit of analysis is an individual observation or case, but in MR-QAP, the unit of analysis is a dyad, which is a pair of individuals or entities that may or may not have some sort of relationship connecting them. In MR-QAP analyses, the dependent variable is a measure of the relationship between two nodes in a network, such as the presence or absence of a tie, the strength of a tie, or the similarity between two nodes. The independent variables are usually node-level attributes, such as personal or behavioral characteristics. The MR-QAP regression can be expressed formally as a linear model, wherein the tie random variables Y_{ij} serve as the dependent variables, while the explanatory variables Z represent the structural positions of network mechanisms. These structural positions are determined by actor covariates of the sender and receiver, dyadic covariates, or other relational factors among the actors. The overall model can be represented in the following general form:

$$g\left(Y_{ij}\right) = \beta_0 + \beta_1 X_1 + \beta_2 X_2 + \cdots + \beta_n X_n \tag{5.15}$$

The edge function g plays a crucial role. When dealing with a valued network, where the edges can have continuous or numeric values, the edge function g simplifies to the identity function. Consequently, the MR-QAP regression assumes the shape of a traditional linear regression model, accommodating continuous predictors. However, in the case of a binary network, where the edges are limited to binary values (0 or 1), the edge function g transforms into the logit function. This modification results in the MR-QAP regression adopting the structure of a logistic regression model, enabling analysis and prediction of binary outcomes.

MR-QAP estimates the coefficients of the independent variables and their significance while controlling for the interdependence between the observations in the network. MR-QAP works by comparing the observed values of the dependent variable with the values that would be expected if the independent variables had no effect on the dependent variable. The expected values are obtained by randomly permuting the independent variables across the nodes while keeping the network structure fixed. This process is repeated many times, and the average difference between the observed and expected values is calculated. The regression coefficients are then estimated by minimizing the sum of the squared differences between the observed and expected values (Krackardt, 1987; Dekker et al., 2007). The permutation has several properties, and they refer almost exclusively to its probability distribution. Specifically, permutation invariance of a distribution means that the distribution is unaffected by any permutation of its elements. In other words, if we have a set of data, we can shuffle or rearrange its order in any way and the resulting distribution of the data will be the same. This property is important because it allows us to use permutation tests and non-parametric methods to analyze data without making assumptions about the underlying distribution of the data. For example, suppose we have a dataset of 10 numbers (1, 2, 3, 4, 5, 6, 7, 8, 9, 10). If we randomly shuffle or permute the order of these numbers (e.g., 6, 4, 2, 1, 9, 10, 7, 3, 5, 8), the resulting distribution of the data (e.g., looking through the histogram or density plot) will be the same as the original distribution. In the context of adjacency matrices and graph models, permutation invariance means that permuting the node labels of a graph (i.e., rearranging the order in which the nodes are listed) does not change the probability of the graph. Permutation invariance is an important property in modeling scenarios where the order of the elements does not matter. For example, in the case of graph models, permuting the node labels does not affect the underlying structure of the graph, and therefore the probability of the graph should remain the same. This property ensures that the model is robust to changes in the labeling of the nodes and can generalize to new node configurations (Aldous, 1981). Another property of permutation is its exchangeability. Permutation invariance and exchangeability are closely related concepts, but they are not exactly the same. Permutation invariance refers to the property that the distribution is unaffected by any permutation of its elements, while exchangeability refers to the property that the joint distribution of a set of random variables is invariant under any permutation of the variables. In other words, permutation invariance applies to a single distribution or dataset, while exchangeability applies to the joint distribution of multiple variables. Exchangeability implies permutation invariance, but the reverse is not always true. For example, consider a set of coin tosses. If we assume that the coin is fair, each toss is a random variable with a Bernoulli distribution. If the coin is exchangeable, the joint distribution of the tosses is invariant under any permutation of the tosses. In other words, the order of the tosses does not matter. However, the distribution of each individual toss is not permutation-invariant, since flipping the coin may change the outcome of the toss. In the context of networks, exchangeability refers to the property that the joint distribution of a set of random variables (e.g., the nodes or edges in a network) is invariant under any permutation of the variables. For example, if we have a graph with n nodes, we can represent it as an adjacency matrix A, where A_{ij} is the weight or existence of the edge between nodes i and j. If the nodes in the network are exchangeable, then the joint distribution of the adjacency matrix A is invariant under any permutation of the node labels. In other words, the network structure is the same regardless of the labeling of the nodes (Snijders & Borgatti, 1999).

5.3.1 Stata Code and Output

```
. nwclear

// We set the number of observations
. set obs 70

// The code creates two new variable called gender and age. This assumes that we have set the number of observations to 70 using the set obs command beforehand
. gen gender = (_n > 10) + 1
. gen age = abs(int(rnormal(50,10)))

// The nwhomophily command is used to generate a network that will have a tendency to have more links between individuals of the same gender. Density (0.05): This means that on average, each node will be connected to 0.05*100=5 other nodes. Homophily (5): This means that nodes are 5 times more likely to connect to nodes of the same gender
. nwhomophily gender, density(0.05) homophily(5)

// The nwqap command is used to perform a quadratic assignment procedure (QAP) with two X variables, namely gender and age
. nwqap homophily gender age
```

5.3.2 R Code and Output

```
# Download and recall the packages
> library(sna)
> library(asnipe)

# Using the rgraph() function in sna, generate three random adjacency matrices
> set.seed(1234)
> g1=rgraph(25, m=1, tprob=0.5, mode="graph")
> g2=rgraph(25, m=1, tprob=0.5, mode="graph")
> g3=rgraph(25, m=1, tprob=0.5, mode="graph")

# Test the effect of g2 on g1, controlling for g3
> netlm(g1, g2+g3, mode="graph", nullhyp="qap", test.statistic="t-value")

# Here, we test the effect of g2 on g1 controlling for g3, and effect of g3 on g1, controlling for g2
> mrqap.dsp(g1~g2+g3, directed="undirected")
```

5.4 Bernoulli Graph Models and P Models

A graph is a set of vertices linked or not to one another. Random graphs are graphs in which the connections are random. In particular, they can be used to investigate a system's emergent behavior, or the patterns of behavior that arise from interactions between network nodes (Erdős & Rényi, 1963; Hunter & Handcock, 2006). Several random and stochastic methods have been suggested to study the relationships in a network. One of the earliest models is the Bernoulli random graph model (Gilbert, 1959). For the chance that an edge exists between node *i* and node *j*, the model takes the following form:

$$P\left(Y_{ij}=1\right)=p,\forall i,j\in\{1,\ldots,n\},\ \text{and}\ \ j>i \tag{5.16}$$

with $p \in (0, 1)$. The Bernoulli random graph model is a basic computational model for undirected binary networks. In this model, we assume that the network has n nodes and give a parameter p to represent the likelihood of any two nodes connecting. We assume that the existence or absence of each edge is independent of all other edges. To be more specific, for each possible edge, we select a random number from a Bernoulli distribution with parameter p. If the number is 1, we include the edge in the network; otherwise, we exclude it. This means that the observed network has a probability of

$$P\left(Y_{ij}=y_{ij}\right)=p^{y_+}\left(1-p\right)^{M-y_+}=\frac{e^{\theta y_+}}{k(\eta)} \quad \text{with } y_+=\sum\nolimits_{i,j}y_{i,j} \tag{5.17}$$

where y_+ specifies the size of the network and $M = N(N - 1)$ indicates the number of ties that can exist with a specific amount of vertices. Finally, $k(\eta)$ is the normalizing constant, $k(\eta) = e^{\eta}/(1 + e^{\eta})$. This method is a poor model. When this approach is applied to more efficient models, it yields the so-called p_1 model (Holland & Leinhardt, 1981; Frank & Strauss, 1986; Pattison & Wasserman, 1999; Robins et al., 1999). Compared to the classical model just presented, in this new model, the independence of ties is always assumed but conditioned on the value assumed by a set of network covariates. It is possible to write it as:

$$\begin{aligned}\operatorname{logit}\left[P\left(Y_{ij}=1\right)\right]&=\log\left\{\frac{P\left(Y_{ij}=1\right)}{1-P\left(Y_{ij}=1\right)}\right\}\\&=\theta\left(y_{ij}+y_{ji}\right)+\rho y_{ij}y_{ji}+\alpha_i y_{ij}+\beta_j y_{ji}\end{aligned}\tag{5.18}$$

with θ that indicates the density of the network, ρ is the number of mutual ties (reciprocity), α indicates the out-degree (productivity), and β is the in-degree (attractiveness). The probability mass function of this model is

$$\begin{aligned}P\left(Y_{ij}=y_{ij}\right)&=\frac{\exp\sum_{i\le j}\left\{\theta\left(y_{ij}+y_{ji}\right)+\rho y_{ij}y_{ji}+\alpha_i y_{ij}+\beta_j y_{ji}\right\}}{k\left(\theta,\rho,\alpha,\beta\right)}\\&=\frac{\exp\left(\theta y_{y+}+\rho\sum_{i\le j}y_{ij}y_{ji}\left\{\rho y_{ij}y_{ji}+\sum_i\alpha_i y_{ij}+\sum_j\beta_j y_{+ji}\right\}\right)}{k\left(\theta,\rho,\alpha,\beta\right)}\end{aligned}\tag{5.19}$$

with the denominator $k(\theta,\rho,\alpha,\beta)$ indicating the normalizing constant equal to:

$$\begin{aligned}k\left(\theta,\rho,\alpha,\beta\right)=\Pi_{i\le j}&\left\{+\exp\left(\theta+\alpha_i+\beta_j+\alpha_j+\beta_i+\rho\right)+\right.\\&\left.+\exp\left(\theta+\alpha_i+\beta_j+\alpha_j+\beta_i+\rho\right)\right\}\end{aligned}\tag{5.20}$$

The p_2 model is a variation of the Bernoulli random graph model that adds a degree-dependent component to allow for triad formation to the p_1 model. The p_2 model employs random effects to simplify the way to model the components. In the p_2 model we have

$$\begin{aligned}\operatorname{logit}\left[P\left(Y_{ij}=1\vert\phi\right)\right]&=\log\left\{\frac{P\left(Y_{ij}=1\vert\phi\right)}{1-P\left(Y_{ij}=1\#\phi\right)}\right\}\\&=\theta\left(y_{ij}+y_{ji}\right)+\rho y_{ij}y_{ji}+\alpha_i y_{ij}+\beta_j y_{ji}+\phi_{i,j}\end{aligned}\tag{5.21}$$

$$\text{with}\,\phi=\left(\phi_1,\phi_2,\ldots,\phi_n\right)^T\sim N\left(0,\sigma_\phi^2 Z_n\right)\tag{5.22}$$

where Z_n indicates the nodal effects as nodal random effects ϕ_i $(i = 1, 2, \ldots, n)$. This enables a rather simple solution to the issue of a growing number of parameters with an increasing number of nodes n, as in most instances the nodal effects themselves are not the cause for concern, but only their variance parameter. In comparison to the p_1 model, the p_2 model considers the fact that the formation of triangles (or other higher-order structures) can have a substantial effect on the formation of edges.

Nodes that engage in many triangles, for example, are more likely to be central in the network and have a larger impact on other nodes, which can affect the probability of them forming new edges. Thus, the p_2 model offers a more detailed picture of how edge formation is affected by network configurations. In this regard, the p_2 model considers an essential characteristic of actual social networks, namely the tendency of similar nodes to be connected to one another. The p_2 model, on the other hand, remains a relatively simple model that ignores other key characteristics of social networks, such as the existence of communities or assortative effects. The family of stochastic block models (Nowicki & Snijders, 2001) is an extension of the Bernoulli model and is analogous to the p_2 model. In these models, edges are allocated to latent groups, also known as blocks. In this kind of model, we assume that the network can be divided into K disjoint blocks, with nodes within a block having identical link patterns. Within the network, the units can be conceived of as “communities” or “clusters” (Daudin et al., 2008). In addition to stochastic block models, the idea of latent variables has been extended in so-called latent network models (Krivitsky & Handcock, 2008).

The probability that an edge appears between nodes i and j does not rely on the network structure itself in any of the models described so far. The Exponential Random Graph Model (ERGM), also known as the $p*$ *(p-star)* model and suggested by Frank and Strauss (1986), allows for the inclusion and modeling of this network structure.

5.5 Exponential Random Graph Model (ERGM) for Binary Data

Exponential random graph models (ERGMs), also called $p*$ models, are a useful class of models for representing complex social phenomena. The purpose of an exponential random graph model is to identify the factors that influence the creation of links within a network. They are edge-based models that represent the likelihood of each link as a function of the topology of a specific network and the properties of its nodes (Krivitsky, 2012). As a result, finding the mechanisms influencing link formation is the main goal of an ERGM (Robins et al., 2007). To characterize the observed edge pattern, local structural configurations can be used in combination with nodal or edge variables. This family of models is known as multi-mechanistic network models precisely because they examine numerous network processes at the same time, allowing for an evaluation of their relative significance in describing observed network data. ERGMs define probability distributions over graphs based on the frequency of specific local structural configurations, such as edges, triangles, or stars. Local configurations are so-called because they are assumed to occur in small local contexts within the network. The patterns of a specific current graph are compared to the expected patterns of a random graph. ERGMs are probability models for network structures that use a log-linear model to specify the probability distribution of the network. In an ERGM, the log-odds of a particular network structure are modeled as a linear combination of network statistics, which are summary measures of the network structure. Each network statistic represents a specific feature of the network, such as the number of edges or the presence of triangles. The log-linear model can be expressed as a sum of the network statistics multiplied by their corresponding model parameters, where each parameter represents the effect of a given network statistic on the log-odds of the network structure. The exponential family of distributions is used to specify the probability distribution of the network, with the log-odds of the network structure serving as the natural parameter of the distribution. In this way, an ERGM provides a flexible and interpretable approach for modeling complex network structures.

Considering what has just been said, the likelihood of observing an edge between any pair of nodes *i* and *j* within a network is directly related to a set of independent variables:

$$p\left(Y_{ij}=1\right)=\theta_1 X_1+\theta_2 X_2+\ldots+\theta_n X_n \tag{5.23}$$

Y is the observed network structure and *i* and *j* are two generic nodes. $Y_{ij} = 1$ denotes the presence of a connection between nodes *i* and *j* ($Yij = 0$ represents the absence of a link). θ is a vector of parameters, *X* is a vector of variables. As in the logit model, even in this case there is no assurance that the probability will remain in the range of 0 and 1. As a result, we have to transform a probability into an odds ratio. We can use the following formula:

$$\text{odds}\left(Y_{ij}=1\right)=\frac{p\left(Y_{ij}=1\right)}{1-p\left(Y_{ij}=1\right)}=\frac{p\left(Y_{ij}=1\right)}{p\left(Y_{ij}=0\right)} \tag{5.24}$$

where *p* is the probability of a single link in the graph *Y*. This odds ratio gives us the likelihood of the edge occurring (*p*) compared to the likelihood of the link not occurring ($1 - p$).

The formula ensures that the odds ratio is a meaningful and interpretable measure of the likelihood of an event occurring and that it satisfies certain mathematical properties. If the probability of an event is 0 (implying that the event is impossible), then the odds ratio obtained from the formula will also be 0, indicating that the event has zero chance of occurring. If the probability of an event is 1 (implying that the event is certain to occur), then the odds ratio obtained from the formula will be infinity, indicating that the event has an infinitely high chance of occurring compared to the chance of it not occurring. The natural logarithm is applied to the odds ratio to produce the log-odds, which can take any real value and are used in the linear predictor of the logistic model. The inverse of this function ensures that the predicted probability remains between 0 and 1:

$$\log\left(\text{odds}\left(Y_{ij}=1\right)\right)=\text{logit}\left(p\left(Y_{ij}=1\right)\right)=\log\left(\frac{p\left(Y_{ij}=1\right)}{p\left(Y_{ij}=0\right)}\right) \tag{5.25}$$

The left-hand side of the equation represents the logarithm of the odds ratio of a specific event occurring, namely the event of a single edge Y_{ij} being present in graph Y. The right-hand side of the equation involves the logit function, which is used to convert the probability of the edge occurring into a log-odds scale, where the odds are the ratio of the probability of the edge occurring to the probability of it not occurring. Specifically, $logit(p(Y_{ij} = 1))$ represents the log-odds of the edge occurring. The final term, $log(p(Y_{ij} = 1)/p(Y_{ij} = 0))$, represents the logarithm of the ratio of the probability of the edge occurring to the probability of it not occurring. This term corresponds to the logarithm of the odds of the edge occurring, that is, the logit of the probability, which is a commonly used quantity in logistic regression models. If we assume that the probability of an edge in a graph is a function of a vector of nnn variables and their corresponding parameters $(\theta 1, \theta 2, \ldots, \theta n)$, we can write the probability as:

$$p\left(Y_{ij} = 1 \middle| X\right) = f\left(X, \theta\right) \tag{5.26}$$

where $X = (X_1, X_2, \ldots, Xn)$ is a vector of n variables, and f is a function that maps X and parameters θ to the probability that the edge $(Y_{ij} = 1)$ is present. The notation $p(Y_{ij} = 1 \mid X)$ means the conditional probability of Y_{ij} being equal to 1 (i.e., the edge is present) given the values of the variables X. The logit of the probability is

$$\operatorname{logit}\left(p\left(Y_{ij} = 1\right)\right) = \theta_1 X_1 + \theta_2 X_2 + \cdots + \theta_n X_n \tag{5.27}$$

$$p\left(Y_{ij} = 1\right) = \exp\left\{\theta_1 X_1 + \theta_2 X_2 + \cdots + \theta_n X_n\right\} \tag{5.28}$$

The equation presented up to this point corresponds to the standard logistic regression model, where the probability of a tie between i and j is assumed to be independent of the probability of a tie between i and k. As we have said before, ERGMs, unlike the logistic model, add the assumption of dependency. The modification introduced by Strauss and Ikeda (1990) consists of adding a term to the logistic regression equation that accounts for the influence of the network structure on the probability of a tie between two nodes. Specifically, the modified equation is:

$$\operatorname{odds}\left(Y_{ij} = 1\right) = \frac{p\left(Y_{ij} = 1 | Y_{ij}^c\right)}{1 - p\left(Y_{ij} = 1 | Y_{ij}^c\right)} = \frac{p\left(Y_{ij} = 1 | Y_{ij}^c\right)}{p\left(Y_{ij} = 0 | Y_{ij}^c\right)} \tag{5.29}$$

In this Eq. (5.29), the odds of an edge between vertices i and j are not exclusively determined by the characteristics of vertices i and j themselves, but also depend on the structure of the network before the tie is created (denoted by Y_{ij}^c). This means that the likelihood of a tie between nodes i and j is conditional on the existing ties in the network, which can affect the likelihood of forming a new tie between these vertices.

We previously discussed how some model variables may be subgraphs. In some network models, subgraphs can be included as variables by simply counting the number of times a given subgraph appears in the network. When subgraphs are used as variables in a network model, the counts of these variables can differ depending on whether a link between two nodes is present or absent. For example, suppose we are using the count of triangles as a variable in a network model. If a link is added to the network, the count of triangles may increase by a certain amount. However, if a link is removed from the network, the count of triangles may decrease by a different amount. To account for the differences in variable values when a link is present versus when it is absent, we can use two vectors of variables: $\upsilon\left(Y_{ij}^{+}\right)$ and $\upsilon\left(Y_{ij}^{-}\right)$. The vector $\upsilon\left(Y_{ij}^{+}\right)$ represents the values of the subgraph variables when a link is added between nodes *i* and *j*. Conversely, the vector $\upsilon\left(Y_{ij}^{-}\right)$ represents the values of the subgraph variables when the link between nodes *i* and *j* is absent. By using both $\upsilon\left(Y_{ij}^{+}\right)$ and $\upsilon\left(Y_{ij}^{-}\right)$ in the network model, we can account for the effects of subgraph variables in different network states and develop more accurate models of network growth and evolution. Considering that, we may rewrite previous Eq. 5.29 by including this differentiation and using the results from Eq. 5.28:

$$\text{odds}\left(Y_{ij}=1\right)=\frac{p\left(Y_{ij}=1 \mid Y_{ij}^{c}\right)}{p\left(Y_{ij}=0 \mid Y_{ij}^{c}\right)}=\frac{\exp\left\{\theta'\cdot\upsilon\left(Y_{ij}^{+}\right)\right\}}{\exp\left\{\theta'\cdot\upsilon\left(Y_{ij}^{-}\right)\right\}} \tag{5.30}$$

With $\upsilon\left(Y_{ij}^{+}\right)$ that represents the vector of variables in the network when the link between the generic *i* and *j* is present and $\upsilon\left(Y_{ij}^{-}\right)$ represents the vector of variables when the link between the generic *i* and *j* is not present. We can further develop the previous Eq. (5.30) as follows:

$$\begin{aligned}\frac{\exp\left\{\theta'\cdot\upsilon\left(Y_{ij}^{+}\right)\right\}}{\exp\left\{\theta'\cdot\upsilon\left(Y_{ij}^{-}\right)\right\}}&=\exp\left\{\theta'\cdot\upsilon\left(Y_{ij}^{+}\right)\right\}\cdot\exp\left\{-\theta'\cdot\upsilon\left(Y_{ij}^{-}\right)\right\}\\&=\exp\left\{\theta'\left[\upsilon\left(Y_{ij}^{+}\right)-\upsilon\left(Y_{ij}^{-}\right)\right]\right\}\end{aligned} \tag{5.31}$$

where $\left[\upsilon\left(Y_{ij}^{+}\right)-\upsilon\left(Y_{ij}^{-}\right)\right]$ is the variation in the count numbers of network statistics that would result from the addition of a single link in the network. Therefore, the parameters of the model are linked not with the counts of sub-structures themselves (e.g., the number of triangles), but rather with the difference in counts between the observed network and a hypothetical network that is identical except for the presence of a single additional link. As a result, the endogenous variables in an ERGM are known as "change statistics". When we have a set of variables and related parameters, and we use the logarithm to remove the exponential, we can formulate the above Eq. (5.31) as follows:

$$\log\left(\frac{\exp\left\{\theta'\cdot\upsilon\left(\mathrm{Y}_{\mathrm{ij}}^{+}\right)\right\}}{\exp\left\{\theta'\cdot\upsilon\left(\mathrm{Y}_{\mathrm{ij}}\right)\right\}}\right)=\theta_1'\left[\upsilon_1\left(\mathrm{Y}_{\mathrm{ij}}^{+}\right)-\upsilon_1\left(\mathrm{Y}_{\mathrm{ij}}\right)\right]+\theta_2'\left[\upsilon_2\left(\mathrm{Y}_{\mathrm{ij}}^{+}\right)-\upsilon_2\left(\mathrm{Y}_{\mathrm{ij}}\right)\right]+\ldots+$$
$$+\theta_n'\left[\upsilon_n\left(\mathrm{Y}_{\mathrm{ij}}^{+}\right)-\upsilon_n\left(\mathrm{Y}_{\mathrm{ij}}\right)\right]= \tag{5.32}$$
$$=\theta_1'\left[\upsilon_1\left(\Delta_1 Y_{ij}\right)\right]+\theta_2'\left[\upsilon_2\left(\Delta_2 Y_{ij}\right)\right]+\ldots+\theta_n'\left[\upsilon_n\left(\Delta_n Y_{ij}\right)\right]$$

with $\Delta_1 Y_{ij}$ the change statistic for a link between i and j for a specific variable. From Eq. 5.32, we notice that the probability of observing a particular network is proportional to the exponential of a linear combination of network statistics or features. As with logistic regression (Besag, 1972), the Hammersley–Clifford theorem provides a justification for why this approach works. Specifically, it shows that any distribution over a set of graphs that satisfies certain properties (including positivity) can be expressed as a Markov random field over a set of binary variables. This means that the probability distribution over the graph can be written in terms of a set of possible functions that depend on the values of these binary variables (corresponding to the edges of the graph). ERGMs are a special case of this more general framework, where the possible functions are chosen to be exponential functions of linear combinations of the graph statistics. By using the Hammersley–Clifford theorem, we can show that these models are consistent with certain assumptions about the underlying structure of social networks, and we can use them to make inferences about the processes that generate observed network data. More specifically, the theorem states that any distribution over a set of graphs that satisfies the following two properties is a Markov random field:

(i) **Positivity**: The probability distribution assigns non-zero probability to all possible graph configurations.
(ii) **Local Markov property**: The probability of a graph Y only depends on the values of a set of binary variables (one for each edge in the graph) that are "local" to Y, meaning that they only depend on the immediate neighbors of each edge.

The Hammersley–Clifford theorem is important because it provides a way to represent complex distributions over graphs using simple graphical models, which can be more tractable for inference and analysis (Besag, 1972; Anderson et al., 1999). Suppose we want to estimate the probability of a triangle in a network using an ERGM. A triangle is a sub-structure in a network that consists of three nodes,

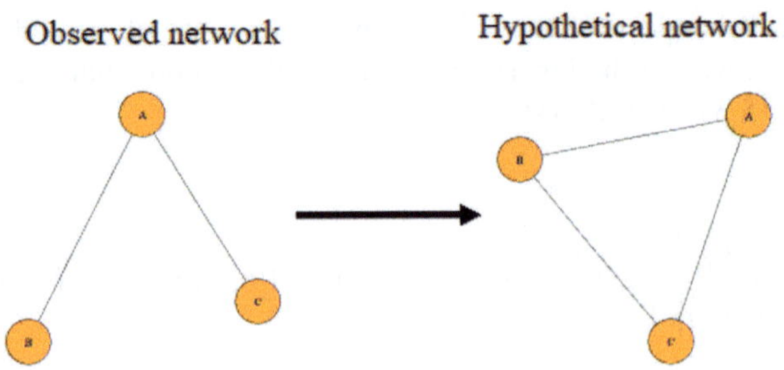

Fig. 5.1 Observed network and hypothetical network

where each node is connected to the other two nodes. To capture the differences in the number of triangles between the observed network and a hypothetical network, we use change statistics. Specifically, we might use a change statistic that counts the number of triangles in the observed network, as well as the number of triangles that would be present in a hypothetical network that differs from the observed network by the presence or absence of a single edge. This way, the ERGM framework captures the dependence structure between different sub-structures in the network. In this example (Fig. 5.1), we have our observed network composed of three nodes with links between A and B and A and C. The edge inserted between B and C results in the formation of a triangle.

We use a change statistic that counts the number of triangles in each network and compares the number of triangles between the observed network and the hypothetical network that differs from the observed network by the presence (or absence) of a single edge. In our hypothetical network, we have added a single edge to the observed network (the edge between B and C). We then examine how this change affects the number of triangles in the network and, consequently, how it affects the likelihood of the network under the ERGM. By comparing the triangle counts before and after the edge modification, we can estimate the contribution of that edge to the probability of observing the overall network structure. Following is an example implementation of the ERGM algorithm in Python using the *NetworkX* and *numpy* packages. To simplify, the following approach just uses the number of edges as a change statistic.

5.5.1 Python Code and Output

```
# Importing the required packages
>>> import numpy as np
>>> import networkx as nx

# Define the change statistic function for the ERGM
>>> def change_statistic(net):
# Count the number of edges in the network
>>>     return np.sum(nx.to_numpy_array(net))

# Define the log-likelihood function for the ERGM
>>> def log_likelihood(net, params):
# Calculate the expected number of edges
>>>     expected_edges = np.exp(params[0] * change_statistic(net))

# Calculate the log-likelihood
>>>     ll = np.sum(np.log(expected_edges) - expected_edges)
>>>     return ll

# Define the function to simulate a new network from the ERGM
>>> def simulate_ergm(params):
# Initialize an empty network
>>>     net = nx.Graph()
# Add nodes to the network
>>>     net.add_nodes_from(range(10))

# Generate edges according to the ERGM
>>>     for i in range(10):
>>>         for j in range(i + 1, 10):
# Calculate the probability of an edge between nodes i and j
>>>             prob_edge = np.exp(params[0])

# Add an edge between nodes i and j with probability prob_edge
>>>             if np.random.random() < prob_edge:
>>>                 net.add_edge(i, j)

>>>     return net

# Generate a random network and fit an ERGM to it
>>> net = nx.erdos_renyi_graph(10, 0.3)
>>> params = np.array([-2.0])
>>> ll = log_likelihood(net, params)
>>> print(f"Initial log-likelihood: {ll:.2f}")

# Simulate a new network from the ERGM and calculate its log-likelihood
>>> new_net = simulate_ergm(params)
>>> new_ll = log_likelihood(new_net, params)
>>> print(f"Simulated log-likelihood: {new_ll:.2f}")
```

This code first defines a function *change_statistic()* that counts the number of edges in a network. The *log_likelihood()* function takes a network and a vector of parameters as input and calculates the log-likelihood of the ERGM. The *simulate_ergm()* function generates a new network from the ERGM by first adding nodes to the network and then adding edges between pairs of nodes according to the probability specified by the ERGM. Finally, the code generates a random network using the *nx.erdos_renyi_graph()* function from the *NetworkX* package, fits an ERGM to it with the specified parameter values, and calculates the initial log-likelihood. It then generates a new network from the ERGM using the *simulate_ergm()* function and calculates its log-likelihood under the same parameter values.

5.5.2 The Probability Distribution of ERGMs

The probability distribution of a network is a function that, given a collection of vertices and edges, provides a probability to each possible network topology (Zinilli et al., 2024b). In other words, given the underlying probability model, it describes the likelihood of observing a specific network configuration. The specific network configuration is the observed network. The observed network is considered to be the result of some (unknown) random process and is one realization from a collection of alternative networks with identical key properties (at the very least, the same number of vertices). In other words, the observed network is viewed as a single edge pattern among many other possible patterns.

When we use a network model to describe the process by which a network is formed, the range of possible networks and their probability of occurrence are represented by a probability distribution on the set of all possible graphs with a given number of nodes. The ability to generate random graphs from the probability distribution allows us to compare the properties of the sampled graphs to the observed network and evaluate whether the model is a good fit for the data. This can help us gain insights into the underlying processes that shape the structure of the network and identify key features that distinguish the observed network from networks generated by chance.

For example, in the Erdős–Rényi model, the number of edges follows a binomial distribution, which gives the probability of observing each possible number of edges in the network, given the number of nodes and the edge probability. The process of network formation is not entirely random but rather influenced by the patterns of connectivity and attributes of the nodes in the network. This can result in the formation of certain structural patterns that are not present in a purely random network (e.g., transitivity).

In Exponential Random Graph Models (ERGMs), each network tie is typically treated as a random variable with a certain probability distribution that reflects the likelihood of observing that edge in the network. The probability distribution for each edge can be influenced by various factors, such as the presence or absence of other edges and node-level attributes. By incorporating these factors into a probabilistic model of network formation, we can generate synthetic networks that reflect the observed patterns of connectivity and provide insights into the underlying mechanisms that drive network formation. Once the probability distribution for each edge has been specified, we can use statistical methods to estimate the parameters of the model from the observed network data. Exponential random graph models have the following form:

$$P_\theta\left(Y = y\right) = \frac{\exp\left(\theta\, g\left(y\right)\right)}{\kappa\left(\theta\right)} \tag{5.33}$$

where $P_\theta(Y = y)$ is the probability of observing a network Y that has a certain set of edges y, θ is a vector of parameters that determine the strength of the effects of the structural features, *g(x)* is a vector of sufficient statistics that describe the structural features of the network (in the next paragraph, we show some of these statistics), and $\kappa(\theta)$ is a normalizing constant that ensures that the probabilities sum to one over all possible graphs with the same number of nodes:

$$\kappa(\theta) = \log Z(\theta;Y) = \log \sum_{y \in Y} \exp(\theta\, g(y)) \tag{5.34}$$

As conditional form we have

$$\text{logit}\left[P\left(Y_{ij} = 1 \middle| Y_{k,l}, (k,l) \neq (i,j); \theta\right)\right] = \theta^T g_{ij}(y) \tag{5.35}$$

Taking into account the difference in counts of change statistics as explained in the previous paragraph, we have

$$\begin{aligned} g_{ij}(y) = \Delta_{ij} g(y) = g\left(y_{ij} = 1, y_{kl}, (k,l) \neq (i,j)\right) \\ -g\left(y_{ij} = 0, y_{kl}, (k,l) \neq (i,j)\right) \end{aligned} \tag{5.36}$$

The counts of change statistics are used to estimate the model parameters and the normalizing constant, which are then used to calculate the probabilities of different graphs. The assumption behind this approach is that the distribution of change statistics in the observed data is indicative of the distribution of change statistics in the population of interest. Simulation methods are used to estimate the probabilities of different graphs under an ERGM because exact computation of these probabilities is often infeasible due to the large number of possible graphs, even for relatively small networks.

Considering the normalizing constant, the number of possible graphs with n nodes $2\binom{n}{2}$, which expands extremely quickly as n increases. For instance, there are $(5 \times 4)/2 = 10$ possible dyads with 5 nodes in an undirected graph, each of which can have a value of 0 or 1. Therefore, there are $2^{10} = 1024$ graphs that must be counted to compute analytically the normalizing constant $\kappa(\theta)$.

In simulation-based approaches, graphs are generated by repeatedly updating the network according to the rules of the ERGM until a large enough sample of graphs is obtained. The frequency of each graph in the sample is then used to estimate probabilities over statistics. Simulation methods can be computationally intensive, especially for large networks or models with many parameters. To address this challenge, various techniques have been developed to improve the efficiency of simulation-based estimation, such as parallel computing and Bayesian inference. These techniques can help reduce the computational burden of fitting ERGMs to large network data and make it possible to explore more complex models that capture more nuanced patterns of dependence among network ties.

The log normalizing constant (also known as the log partition function) is derived from the features of the exponential family type distribution:

$$\begin{aligned}\frac{\partial}{\partial\theta}\log\left(\kappa(\theta)\right) = \mathrm{E}\left(g(x)\theta\right) \xrightarrow{\text{and}} \frac{\partial^2}{\partial\theta^T}\log\left(\kappa(\theta)\right) \\ = \operatorname{cov}\left(g(x)^T \mid \theta\right)\end{aligned} \tag{5.37}$$

where θ^T refers the transposed vector of parameters associated with the sufficient statistics of the network. One of the properties of exponential family type distributions is that the log partition function can be expressed as the logarithm of a summation over the exponential of the inner product between the vector of sufficient statistics and the parameters. This summation is taken over all possible values of the sufficient statistics, which in the case of ERGMs is the set of all possible graphs with the same number of nodes.

5.5.3 *ERGM as Markovian Model*

Exponential random graph models are a type of Markovian model used in network analysis. In this context, the term “Markovian” refers both to the local dependence assumption and to the simulation or estimation phase. The notion of local dependence aligns with the Markov property as formalized in the Hammersley–Clifford theorem, which justifies representing the global probability distribution of a network through local configurations. Regarding the simulation or estimation phase, a Markov chain is used to explore the space of possible network configurations. This Markovian property becomes evident during simulation procedures—such as those based on Markov Chain Monte Carlo (MCMC) methods—where the current network configuration at iteration t depends only on the configuration at iteration $t - 1$. In this setting, the sequence of sampled networks forms a Markov chain that converges to the target ERGM distribution. This makes ERGMs a type of discrete-time Markov chain model, where the possible states of the system are the different possible network configurations (i.e., graphs) and the transition probabilities between these states are given by the ERGM. The Markovian nature of ERGMs is also reflected in the use of change statistics, which capture the local changes in the network structure that occur when a single edge is added or deleted. Specifically, the Markovian property of ERGMs means that the probability of a particular network structure at time t depends only on the previous network structure at time $t - 1$ and the change statistics associated with the edge that was added or deleted. This makes change statistics a natural way to parameterize the model and capture the underlying network dynamics. In a few words, the change statistics (for instance, the number of triangles) allow the model to account for the dependencies between the different network configurations that arise from the Markovian nature of the process.

Exponential Random Graph Models (ERGMs) and other models for network analysis can suffer from degeneracy and instability problems. Degeneracy occurs when the model places a disproportionate amount of probability mass on only a few of the possible graph configurations in the space of possible graphs Y. For example, a model might assign a very high probability to only empty or complete graphs, while assigning low probabilities to all other graph configurations. This makes the model less useful for analyzing real-world networks, which are typically much more complex and diverse than these simple configurations. Instabilities are another problem that can occur in random graph models. A model is said to be unstable if small changes in the parameter values result in large changes in the probabilistic structure of the model. This means that the model is not robust and can produce very different results depending on the precise values of the parameters chosen. This makes it difficult to interpret and use the model for real-world applications, as small changes in the data or model assumptions can lead to unpredictable changes in the results. Researchers have proposed various methods for dealing with degeneracy and instability in ERGMs and other random graph models, such as using model selection techniques to choose the most appropriate model for a given network, or developing more complex models that can capture a wider range of graph configurations (Robins et al., 2007). One such approach is the partial dependence concept, which was introduced by Pattison and Robins (2002). This concept involves specifying the model in terms of the number and type of partial sub-structures (such as triads or stars) that occur in the network, rather than specifying the entire dependence structure. Alternating k-statistics have been found to reduce degeneracy and instability problems while also improving the models' ability to capture specific sub-configurations (Snijders et al., 2006).

5.5.4 Sufficient Statistics

One of the advantages of ERGMs is the variety of sufficient statistics or terms that may be used to describe network behaviors. The so-called sufficient statistics refer to a set of summary measures of a network structure that capture all of the relevant information about that structure for the purposes of estimating the model parameters (Handcock et al., 2008, 2015). The Exponential random graph model is a statistical method that calculates the probability of a network structure given a set of node-level and dyad-level covariates, and the likelihood function used to estimate the model parameters can be computationally intensive due to the large number of possible network structures. However, by identifying a set of sufficient statistics that capture the most important features of the network structure, the computational problem can be greatly reduced. Sufficient statistics in ERGMs are typically functions of graph counts, which are counts of small, connected subgraphs in the network, such as triangles or four-node cycles. By using graph counts as sufficient statistics, the likelihood function can be expressed based on these counts, rather than the entire network structure, which greatly simplifies the estimation process. Furthermore, the use of sufficient statistics can also improve the interpretability of the model results, as the model parameters can be directly linked to specific features of the network structure (Snijders et al., 2006). The choice of sufficient statistics in an ERGM depends on the specific research question and the network being studied. In general, the goal is to choose sufficient statistics that capture the important patterns of ties in the network. Some commonly used sufficient statistics in ERGMs include:

- *Edge counts*. The number of edges in the network is a simple statistic that can be used to control for the overall density of the network.
- *Triad census*. The triad census is a count of the number of occurrences of each possible three-node subgraph in the network. Triads are important because they capture the tendency for networks to exhibit clustering or transitivity.
- *Alternating k-stars*. A *k-star* is a subgraph in which one node is connected to *k* other nodes, and those *k* other nodes are not connected to each other. *K-stars* are important in ERGMs because they capture the tendency for nodes to be connected to other nodes in star-like patterns, which can be indicative of hierarchy or power relationships in the network.
- *Geometrically weighted degree (GWDegree)*. The geometrically weighted degree (*GWDegree*) is a measure of the tendency for nodes to be connected to other nodes that are similar in terms of their degree, but not necessarily identical. The *GWD* is a more flexible measure than the degree distribution, as it allows for a wider range of possible degree assortative patterns. The geometrically weighted degree (*GWDegree*) is calculated using the formula:

$$d(y,\alpha) = e^{\alpha} \sum_{i=1}^{n-1} \left\{ 1 - \left(1 - e^{-\alpha}\right)^{i} \right\} D_i(y) \tag{5.38}$$

where α represents the geometric decay rate that determines the strength of the geometric function applied to node degree. The contribution of higher degree nodes is considerably reduced for large values of α. As the weight parameter α approaches 0, the geometric function in the *GWD* statistic becomes less steep, which means that edges with larger degrees are given increasingly more weight (Snijders et al., 2006).

- *Geometrically weighted edge-wise shared partners* (*GWESP*). It is based on the idea that two nodes in a network are more likely to be connected if they share a specific number of neighbors. The measure is computed by assigning decreasing weights to higher numbers of shared partners, using a geometric decay function.

$$w(y,\alpha) = e^{\alpha} \sum_{i=1}^{n-2} \left\{ 1 - \left(1 - e^{-\alpha}\right)^{i} \right\} E_i(y) \tag{5.39}$$

The decay parameter α controls the rate at which the weights decline as the number of sharing partners increases. According to Hunter (2007), larger values of the decay parameter result in slower decay. This means that the influence of having more shared partners decreases more gradually as their number increases. However, it is important to note that the estimation of decay parameters for the *GWESP* statistic can be done using methods described in the curved exponential family models of Hunter and Handcock (2006). In practice, in most cases, the decay parameters are assumed to be fixed, meaning they are predetermined instead of being estimated from data.

- *Geometrically weighted dyad-wise shared partners* (*GWDSP*). It is a network statistic that quantifies the number of nodes in a network that share common neighbors. It is similar to the GWESP statistic in that it applies geometrically decreasing weights to the number of shared partners, but differs in that it considers all dyad pairs, regardless of whether an edge exists between them. The geometrically weighted dyad-wise shared partners (*GWDSP*) is calculated using the formula:

$$l(y,\alpha) = e^{\alpha} \sum_{i=1}^{n-2} \left\{1 - \left(1 - e^{-\alpha}\right)^{i}\right\} D_i(y) \tag{5.40}$$

where α indicates the weight given to nodes that share more common neighbors in terms of dyads (pairs of nodes). Higher values of α indicate a stronger weight for nodes that share more dyad-wise common neighbors. In general, the choice of the weight parameter α in the geometric function depends on the specific application. The value is nonnegative, and the values of α between 0.1 and 0.5 are often used.

5.6 Estimation of ERGMs

The estimation of Exponential Random Graph Models (ERGMs) involves the process of determining the coefficients or parameters that best fit the model to the observed network data. Maximum likelihood estimation (MLE) is a commonly used approach for estimating the parameters of ERGMs (Yon et al., 2021). The basic idea behind MLE is to find the parameter values that maximize the likelihood function of the model given the observed data. The likelihood function describes how likely the observed data are given the parameter values of the model (Snijders, 2002). The goal of MLE is to find the parameter values that make the observed data most likely. The likelihood function for an ERGM is often expressed as a product of the probabilities of each possible network structure, where each probability is a function of the network statistics and the model parameters. As mentioned before, the network statistics are local features of the network, such as the number of ties between pairs of nodes, and are used to capture the network structure. To estimate the model parameters using MLE, one needs to find the maximum of the likelihood function.

$$\ell(\theta) = \theta^{T} g(y) - \log \kappa(\theta, \Upsilon) \tag{5.41}$$

where the observed network is denoted as y. Instead of directly maximizing the function $\ell(\theta)$, we will instead focus on evaluating the log-ratio of likelihood values.

$$\ell(\theta) - \ell(\theta_0) = (\theta - \theta_0)^{T} g(y) - \log\left[\frac{\kappa(\theta, \Upsilon)}{\kappa(\theta_0, \Upsilon)}\right] \tag{5.42}$$

Here, θ_0 represents an arbitrarily selected parameter vector. The task of approximating ratios of normalizing constants, as exemplified in expression (5.42), is a complex and extensively explored problem in the literature. Geyer and Thompson (1992) introduced an important concept that forms the core of the ERGM function. This concept, derived through algebraic manipulations of equations in (5.33) and (5.34), serves as a fundamental idea that we extensively leverage.

$$\left[\frac{\kappa(\theta,\Upsilon)}{\kappa(\theta_0,\Upsilon)}\right] = E\theta_0 \exp\left\{(\theta-\theta_0)^T g(Y)\right\} \tag{5.43}$$

By denoting $E\theta_0$ as the expectation under the assumption that Y follows the distribution given by $P\theta_{0,\Upsilon}$, we can exploit the law of large numbers to our advantage. This allows us to approximate the log-ratio by capitalizing on the convergence of sample averages to their expected values.

$$\ell(\theta)-\ell(\theta_0) \approx (\theta-\theta_0)^T g(y) - \log\left[\frac{1}{m}\sum_{i=1}^{m}\exp\left\{(\theta-\theta_0)^T g(Y_i)\right\}\right] \tag{5.44}$$

where $Y_1, Y_2, Y_3, \ldots, Y_m$ is a random sample from the distribution defined by $P\theta_{0,\Upsilon}$, using a Monte Carlo Markov Chain (MCMC) routine, as outlined in the next section (Hunter et al., 2008a). This simulation technique enables us to obtain representative samples from the target distribution for further analysis and approximation purposes. The stochastic estimation technique discussed above relies on selecting a parameter value θ_0. While Eq. 5.44 can theoretically provide precise approximations by increasing the sample size of the Monte Carlo Markov Chain (MCMC), in practice, achieving accurate results can be challenging unless θ_0 is carefully chosen. If θ_0 is not chosen properly, Eq. 5.44 will not produce reliable outcomes in terms of accurately estimating the maximum likelihood estimator. Therefore, selecting an appropriate θ_0 is crucial to ensure the reliability of the approximation results. As a result, the choice of θ_0 plays a crucial role in the stochastic algorithm for approximating the MLE. It has to be "close enough" to the true MLE estimate in order to ensure the effectiveness and reliability of the approximation method. This can be achieved using numerical optimization techniques, such as the Newton–Raphson method, or more commonly, iterative Markov Chain Monte Carlo (MCMC) simulations. In the MCMC approach, a Markov chain is generated with each step proposing a new network structure consistent with the current model parameters. The Markov chain converges to the distribution defined by the model parameters, which allows for the approximation of the normalizing constant and likelihood function in ERGM estimation. Once the MCMC algorithm has converged, the posterior distribution can be used to estimate the model parameters and to perform inference, such as hypothesis testing and confidence interval estimation. In the upcoming sections, the model fit can be evaluated using goodness-of-fit measures. Additionally, we will delve deeper into some of the topics described here and endeavor to provide further clarification.

5.6.1 Markov Chain Monte Carlo (MCMC)

When estimating a network, the process starts with an observed network. The observed network represents the actual network data that we have available for analysis. It can be a binary network where edges represent the presence or absence of connections between nodes, and it can contain additional attributes associated with the edges or nodes. The observed network serves as the basis for estimating the parameters of an ERGM. The goal is to find the set of model parameters that best describe the observed network and capture its underlying patterns and dependencies. The objective in estimating parameter values of an ERGM is to find the parameter values θ that make the observed network the most likely network to have formed among all possible networks according to the distribution of ERGM. The estimate approaches presented in this paragraph will be limited to unweighted networks. Similar work on weighted networks is emerging in the literature (see, for example, Krivitsky, 2012, and Desmarais & Cranmer, 2012). However, research and methodologies for weighted networks are still emerging and are not as well established as those for unweighted networks. MCMC (Markov Chain Monte Carlo) is the most commonly used technique for estimating parameters in ERGMs. MCMC methods are well suited for dealing with the complex dependencies and high-dimensional parameter spaces often encountered in network analysis methods (Snijders, 2002). The advantage of MCMC approaches such as the Metropolis-Hastings algorithm or the Gibbs sampler is that they do not require direct computation of the normalizing constant. Instead, they generate a Markov chain of parameter samples that approximate the posterior distribution of the model parameters given the observed data. This is achieved by iteratively proposing new parameter values and accepting or rejecting them based on an acceptance probability criterion. By executing the MCMC algorithm, a sequence of parameter samples is produced, which gradually converges toward the target distribution. This convergence is typically assessed using diagnostic tools such as trace plots, Gelman–Rubin statistics, or autocorrelation plots. These simulated samples can then be used to estimate the model parameters by finding the values that maximize the likelihood. The power of MCMC lies in its ability to simulate a sample from the target distribution, allowing for efficient exploration of the parameter space and estimation of complex models like ERGMs. MCMC algorithms rely on the Markov property, which states that the probability of transitioning to a future state depends solely on the current state and is independent of the past states. These algorithms propose new parameter values and transition between states in the parameter space based on acceptance probabilities, ensuring that the Markov chain explores the parameter space in a way that gradually approximates the target distribution. Let me quickly remind you of some Markov chain properties. Markov chains are stochastic processes that model situations where the transition between states in a network is not deterministic but probabilistic. They are characterized by the following properties:

- Probabilistic transitions: Markov chains represent transitions between states that occur in a probabilistic manner. The next state is not determined in a deterministic way, but is determined by probabilities associated with each possible transition.
- Homogeneity: The probabilities of transitioning between states in a Markov chain do not depend on the number of transitions that have occurred. This property is known as homogeneity or time-homogeneity, indicating that the transition probabilities remain constant over time.
- Memorylessness: Markov chains exhibit the property of memorylessness, which means that the probability of transitioning to a subsequent state depends solely on the current state and is independent of the past states. The chain does not "remember" the history of how it arrived at the current state.

To construct suitable trajectories of Markov chains from a set of possible states that a Markov chain can take, two common algorithms are used: Gibbs Sampling and the Metropolis–Hastings algorithm. Both algorithms are used for sampling from complex probability distributions, and they can be applied to construct Markov chain trajectories. In the next section, we focus on the Metropolis–Hastings algorithm, which is a commonly used algorithm in the context of Exponential Random Graph Models (ERGMs). The Metropolis–Hastings algorithm allows for the estimation of complex network models and can handle models with a large number of parameters. Furthermore, the Metropolis–Hastings algorithm is a MCMC method, and with appropriate settings and convergence diagnostics, it can provide valid inferences and estimates of model parameters.

5.6.2 *Metropolis–Hastings Algorithm*

Metropolis–Hastings algorithm (Metropolis et al., 1953) is one of the most widely used MCMC methods in ERGM studies. The Metropolis–Hastings algorithm is a Markov Chain Monte Carlo (MCMC) method that can be described as follows:

1. *Initialization:* Start with an initial state $x^{(t)}$ at time $t = 0$.
2. *Proposal step:* Generate a new sample x' from the proposal distribution $Q(x'|x^{(t)})$. Remember that the previously drawn sample $x(t)$ at time step t may have an impact on the proposal distribution.
3. *Acceptance step:* Compute the importance sampling weight ratio, which compares the weights of the new state x' and the current state $x^{(t)}$:

$$\alpha = \frac{W_{x'}}{W_{x^{(t)}}} = \frac{P^*(x')Q\left(x^{(t)}|x'\right)}{P^*\left(x^{(t)}\right)Q\left(x'|x(t)\right)} \quad \text{with } W_{x'} = \frac{P^*(x')}{Q\left(x'|x^{(t)}\right)} \quad \text{and } W_{x^{(t)}} = \frac{P^*\left(x^{(t)}\right)}{Q\left(x^{(t)}|x'\right)} \tag{5.45}$$

where P^* is the unnormalized target distribution. The unnormalized target distribution appears in both the numerator and denominator of Eq. 5.45, allowing the use of the unnormalized distribution without calculating the partition function Z. In the Metropolis–Hastings algorithm, the unnormalized target distribution, denoted as P^*, refers to the target distribution you are trying to sample from without explicitly calculating its partition function (Z). The partition function is a normalization constant that ensures the probability distribution integrates to 1 over the entire sample space. This property is particularly useful in situations where computing the partition function Z is challenging or computationally expensive (as in this case).

Accept the new state with probability:

$$A\left(x' \middle| x^{(t)}\right) = \min\left(1, \alpha\right) \tag{5.46}$$

If accepted, set $x^{(t+1)} = x'$; otherwise, set $x^{(t+1)} = x^{(t)}$. The ratio $A(x'|x^{(t)})$ in the Metropolis–Hastings algorithm represents the probability of transitioning from state x' to state $x^{(t)}$ compared to the probability of transitioning from state $x^{(t)}$ to state x'. This ratio is used to determine whether the new state x' should be accepted or rejected. When the acceptance ratio $A(x'|x^{(t)})$ is higher, it indicates that transitioning from x' to $x^{(t)}$ is more favorable, implying that x' is a good place to move in the sample space. In this case, the algorithm is more likely to accept the new state x', and the Markov chain will move toward x'. Conversely, when the acceptance ratio $A(x'|x^{(t)})$ is low, it suggests that transitioning from $x^{(t)}$ to x' is less favorable. The algorithm tends to stay at the current state $x^{(t)}$, but the proposal distribution Q still allows for exploration in other areas of the sample space. The kernel Q spreads the probability mass, allowing for the possibility of moving to different regions.

4. Repeat steps 2 and 3 until convergence. This process is called the burn-in period and refers to a certain number of initial iterations that are performed to allow the Markov chain to reach a state of approximate convergence. Determining the appropriate length of the burn-in period depends on various factors, including the complexity of the target distribution, the quality of the initial state, and the specific problem at hand. In practice, a commonly used range for the burn-in period is 8.000–12.000 iterations, although this can vary depending on the characteristics of the problem being solved.

The Metropolis–Hastings algorithm captures rare events under Q that have high probability under the target distribution P, leading to a significant shift in the proposal distribution Q. However, it can be challenging to move between high probability regions separated by low probability regions. Although the algorithm will eventually converge to the true distribution P (with certain exceptions), the convergence time is not guaranteed. Convergence in the Metropolis–Hastings algorithm refers to the property where the generated samples progressively approach and approximate the target distribution. Unlike in rejection sampling, where accepted points are independent samples, the samples produced by the Metropolis–Hastings algorithm exhibit dependence on the previous states due to its Markovian nature. Consequently, obtaining effectively independent samples necessitates running the algorithm for a sufficient number of iterations.

To estimate the required number of iterations, we consider the length scales of the state space in both the target and proposal distributions. Let's denote the largest length scale of the target distribution as L and the smallest length scale of the proposal distribution as φ. In this context, a lower bound on the number of iterations, T, can be expressed as:

$$T \approx \left(\frac{L}{\varphi}\right)^2 \tag{5.47}$$

This bound serves as a guideline for selecting an appropriate length scale for the proposal distribution. Adjusting the length scale impacts both the acceptance rate and the level of dependence observed between consecutive samples. In higher-dimensional spaces, the majority of the probability mass is often concentrated within small regions of the state space. Consequently, if the length scale φ is set to be too large, the sampler is more likely to select points from low probability regions, resulting in a higher rejection rate. Conversely, if φ is too small, the sampler will take longer to explore the state space, leading to a slower convergence rate. Determining an optimal value for φ involves selecting a scale that is smaller than L but still on a similar order of magnitude. However, it is crucial to consider the specific properties of the target distribution to ensure an appropriate choice for φ. Monitoring convergence typically involves visualizing the samples obtained from multiple runs of the Metropolis–Hastings algorithm. If the resulting chains exhibit good mixing, meaning the samples are well dispersed and cover the target distribution adequately, it indicates that the samples have likely converged. However, visualizing all dimensions of the random variables becomes challenging in high-dimensional problems. Alternatively, plotting the complete log-likelihood function, which captures the evolution of all model random variables over time, can be useful for detecting convergence. Typically, as convergence is approached, the log-likelihood tends to plateau, providing an indication of convergence.

An example for generating graphs to show the convergence of the Metropolis–Hastings algorithm using R can be found below.

5.6.3 R Code and Output

```
# Function for the Metropolis-Hastings algorithm
> metropolis_hastings <- function(target_distribution, proposal_distribution, initial_state, num_samples) {
  samples <- numeric(num_samples + 1)
  samples[1] <- initial_state
  current_state <- initial_state

  for (i in 1:num_samples) {
    proposed_state <- proposal_distribution(current_state)
    acceptance_ratio <- min(1, target_distribution(proposed_state) / target_distribution(current_state))
    if (runif(1) < acceptance_ratio) {
      current_state <- proposed_state
    }
    samples[i + 1] <- current_state
  }

  return(samples)
}

# Define the target distribution (e.g., a normal distribution)
> target_distribution <- function(x) {
  exp(-x^2)
}

# Define the proposal distribution (e.g., a Gaussian distribution)
> proposal_distribution <- function(x) {
  rnorm(1, mean = x, sd = 0.5)
}

# Set the number of iterations and initial state
> num_iterations <- 5000
> initial_state <- 0

# Generate samples using the Metropolis-Hastings algorithm
> samples <- metropolis_hastings(target_distribution, proposal_distribution, initial_state, num_iterations)

# Plot the samples over iterations
> plot(samples, type = "l", xlab = "Iteration", ylab = "Sample Value", main = "Convergence of Metropolis-Hastings
Algorithm")

# Markers for convergence scenarios
> bad_convergence_marker <- 20
> bad_convergence_color <- "darkred"

# Index for convergence point
> bad_convergence_idx <- which.max(samples)
# Add marker for bad convergence
points(bad_convergence_idx, samples[bad_convergence_idx], pch = bad_convergence_marker, col = bad_convergence_color)
# Add legend
legend("topright", legend = "Bad Convergence", pch = bad_convergence_marker, col = bad_convergence_color)
```

Output

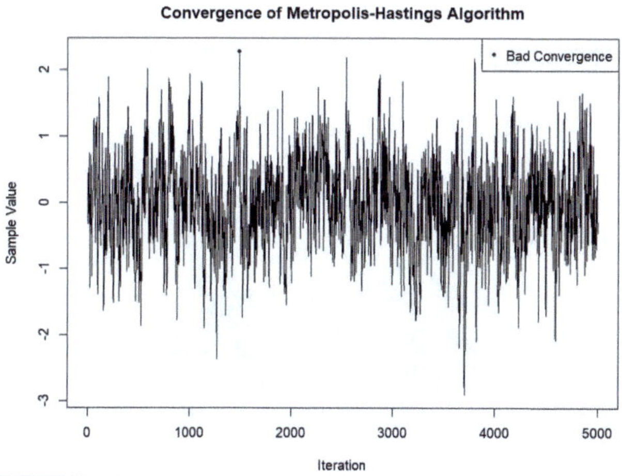

Below you can find an example of generating plots to demonstrate the convergence of the Metropolis–Hastings algorithm using Python and the *Matplotlib* library.

5.6.4 Python Code and Output

```
# Importing the required packages
>>> import numpy as np
>>> import matplotlib.pyplot as plt

# Function for the Metropolis-Hastings algorithm
>>> def metropolis_hastings(target_distribution, proposal_distribution, initial_state, num_samples):
>>>     samples = [initial_state]
>>>     current_state = initial_state

>>>     for _ in range(num_samples):
>>>         proposed_state = proposal_distribution(current_state)
>>>         acceptance_ratio = min(1, target_distribution(proposed_state) / target_distribution(current_state))
>>>         if np.random.uniform(0, 1) < acceptance_ratio:
>>>             current_state = proposed_state
>>>         samples.append(current_state)

>>>     return np.array(samples)

# Define the target distribution (e.g., a normal distribution)
>>> target_distribution = lambda x: np.exp(-x**2)

# Define the proposal distribution (e.g., a Gaussian distribution)
>>> proposal_distribution = lambda x: np.random.normal(x, 0.5)

# Set the number of iterations and initial state
>>> num_iterations = 5000
>>> initial_state = 0

# Generate samples using the Metropolis-Hastings algorithm
>>> samples = metropolis_hastings(target_distribution, proposal_distribution, initial_state, num_iterations)

# Plot the samples over iterations
>>> plt.plot(samples)
>>> plt.xlabel("Iteration")
>>> plt.ylabel("Sample Value")
>>> plt.title("Convergence of Metropolis-Hastings Algorithm")

# Markers for convergence scenarios
>>> bad_convergence_marker = 'o'
>>> bad_convergence_color = 'darkred'

# Indices for convergence points
>>> bad_convergence_idx = np.argmax(samples)

# Add markers for bad convergence
>>> plt.scatter(bad_convergence_idx, samples[bad_convergence_idx], marker=bad_convergence_marker,
color=bad_convergence_color, label="Bad Convergence")

>>> plt.legend()
>>> plt.show()
```

Output

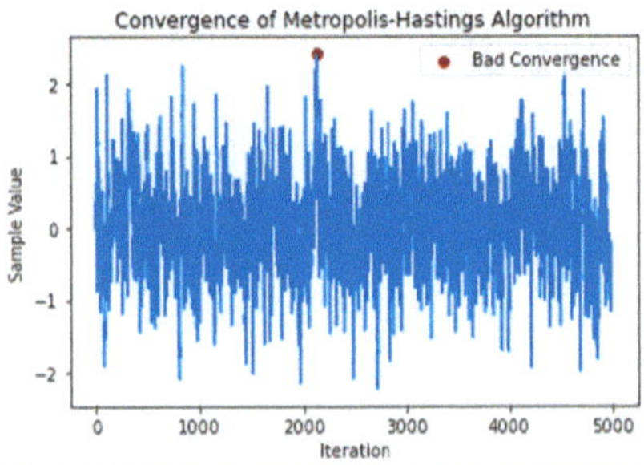

The Metropolis–Hastings MCMC algorithm is used to estimate the parameters of several models, such as linear regression models or ERGMs. It involves defining a likelihood function based on the observed data and a prior distribution representing prior beliefs about the parameters. The algorithm iteratively proposes new parameter values, calculates an acceptance ratio based on the likelihood and prior, and accepts or rejects the proposed values. This process allows the algorithm to explore the posterior distribution and generate samples that represent the uncertainty in the parameter estimates. The accepted samples provide estimates of the parameters, and statistical measures can be computed to quantify their uncertainty. In the next box, an example of how you can apply the Metropolis–Hastings MCMC algorithm to estimate the parameters of a simple linear regression model using Python and the *statsmodels* library.

5.6.5 Python Code and Output

```
# Importing the required packages
>>> import numpy as np
>>> import statsmodels.api as sm
>>> import matplotlib.pyplot as plt

# Generate some simulated data
>>> np.random.seed(0)
>>> X = np.random.randn(100, 2)
>>> X = sm.add_constant(X)
>>> true_params = [1.0, 2.0, 3.0]
>>> epsilon = np.random.randn(100)
>>> y = np.dot(X, true_params) + epsilon

# Define the likelihood function
>>> def likelihood(parameters):
>>>     beta0, beta1, beta2 = parameters
>>>     y_pred = np.dot(X, [beta0, beta1, beta2])
>>>     likelihood = np.sum(-0.5 * np.log(2 * np.pi) - 0.5 * np.square(y - y_pred))
>>>     return likelihood

# Define the prior distribution
>>> def prior():
>>>     return np.random.normal(size=3)

# Define the proposal distribution
>>> def proposal(current_parameters):
>>>     return current_parameters + np.random.normal(size=3)

# Set the number of iterations and burn-in period
>>> num_iterations = 10000
>>> burn_in = 1000

# Initialize the Markov chain
>>> current_parameters = prior()
>>> chain = []

# Run the MCMC algorithm
>>> for i in range(num_iterations):
# Generate new parameter values from the proposal distribution
>>>     new_parameters = proposal(current_parameters)

# Compute the acceptance probability
>>>     acceptance_prob = min(1, np.exp(likelihood(new_parameters) - likelihood(current_parameters)))

# Determine whether to accept or reject the proposed values
>>>     if np.random.uniform() < acceptance_prob:
>>>         current_parameters = new_parameters

# Store the current parameter values in the chain
>>>     if i >= burn_in:
>>>         chain.append(current_parameters)

# Analyze the results
>>> chain = np.array(chain)
>>> mean_parameters = np.mean(chain, axis=0)
>>> interval = np.percentile(chain, [2.5, 97.5], axis=0)

# Plot the parameter trace
>>> fig, ax = plt.subplots(3, 1, figsize=(10, 8))
>>> ax[0].plot(chain[:, 0])
>>> ax[0].set_ylabel('Intercept')
>>> ax[1].plot(chain[:, 1])
>>> ax[1].set_ylabel('Slope 1')
>>> ax[2].plot(chain[:, 2])
>>> ax[2].set_ylabel('Slope 2')
>>> plt.xlabel('Iteration')
>>> plt.tight_layout()
>>> plt.show()

Output
```

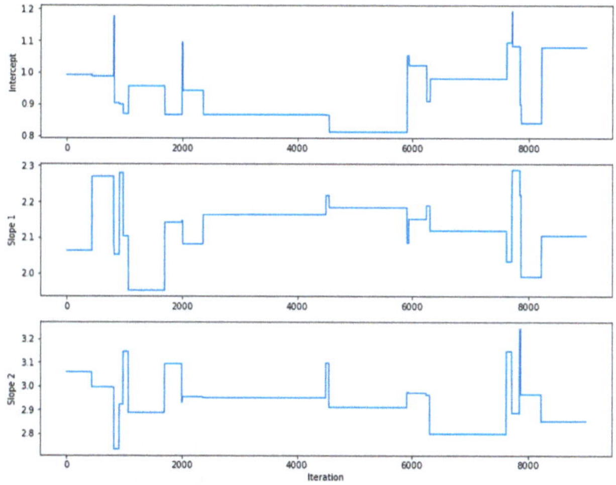


```
>>> print("True Parameters:", true_params)
>>> print("Estimated Parameters (Mean):", mean_parameters)
>>> print("Estimated Parameters (Interval):", interval)
```

In the context of Exponential Random Graph Models (ERGMs), the Metropolis–Hastings algorithm is used to propose transitions from the current network structure y(current) to a proposed network structure y(proposed). The algorithm explores different network structures by making random changes or transitions, such as adding or removing edges, altering the configuration of nodes, or modifying other network characteristics. The goal of the Metropolis–Hastings algorithm is to sample from the posterior distribution of network structures given the observed data. It achieves this by iteratively generating proposed network structures and evaluating their probability according to the ERGM. The proposed network is then accepted or rejected based on a comparison of the probabilities of the current and proposed networks, following the acceptance ratio or auxiliary distribution criteria. By repeatedly proposing and accepting/rejecting transitions, the Metropolis–Hastings algorithm allows for the exploration of different network configurations, eventually converging to the target distribution. These transitions are guided by an auxiliary distribution that depends on the current state $y_{current}$. The purpose of the auxiliary distribution is to either focus the transitions or spread them more broadly throughout the state space Y. The probability of accepting the proposed transition (new graph), in the Metropolis algorithms is denoted as:

$$\min\left\{1, \frac{P\theta_{0,Y}\left(Y = y_{proposed}\right)}{P\theta_{0,Y}\left(Y = y_{current}\right)}\right\} \tag{5.48}$$

where the probability of transitioning to the proposed state is given by the minimum of two values: 1 and the ratio of the probabilities of the proposed network and the current network. In the more general Metropolis–Hastings algorithm, the Eq. 5.48 is replaced by next equation:

$$\min\left\{1, \frac{P\theta_{0,Y}\left(Y = y_{proposed}\right)q\left(y_{current}, y_{proposed}\right)}{P\theta_{0,Y}\left(Y = y_{current}\right)q\left(y_{proposed}, y_{current}\right)}\right\} \tag{5.49}$$

The probability of accepting the proposed transition is modified accordingly. The $q(y_{proposed}, y_{current})$ represents the probability of the proposed state, given that the current state, under the auxiliary distribution. By including both the target and auxiliary distribution probabilities, the acceptance probability in the Metropolis–Hastings algorithm considers the relative likelihoods of the proposed and current networks under both distributions. This allows the algorithm to explore the state space by accepting transitions that improve the fit to the target distribution while taking into account the characteristics of the auxiliary distribution. Metropolis and Metropolis–Hastings algorithms offer a powerful approach for sampling from complex distributions like those in Exponential Random Graph Models (ERGMs). One of the appealing aspects of these algorithms is their ability to avoid the need for computing normalizing constants. In fact, this ratio is

$$\frac{P\theta_{0,Y}\left(Y = y_{proposed}\right)}{P\theta_{0,Y}\left(Y = y_{current}\right)} = \exp\left\{\theta_0\left[g\left(y_{proposed}\right) - g\left(y_{current}\right)\right]\right\} \tag{5.50}$$

If the proposed network state differs from the current state by exactly one edge toggle, meaning that a single edge is flipped from present to absent or vice versa, then the difference in network statistics, $g(y_{proposed}) - g(y_{current})$, can be expressed as $\pm \upsilon_g(y)_{ij}$. Here, $\upsilon_g(y)_{ij}$ represents the change statistic for the specific edge toggle (i,j) in the network. The sign of ± indicates whether the toggle increases or decreases the network statistics. On the other hand, if the proposed network state differs substantially from the current state due to a specific type of Metropolis–Hastings proposal, the ratio in Eq. 5.50 can be calculated by considering a sequence of networks. This sequence starts with the current network and gradually transitions to the proposed network, with each network having one dyad (pair of nodes) different from the previous one. At each step of this sequence, the ratio in Eq. 5.50 can be evaluated by examining the change statistic vector. The change statistic vector represents the incremental changes in the network statistics as we move from one network to the next. By considering the sequence of networks and their corresponding change statistic vectors, the ratio can be computed efficiently without recalculating the entire ratio from scratch (Hunter et al., 2008b).

Metropolis–Hastings algorithms have the capacity to converge more efficiently than other algorithms, such as the Gibbs sampling, specifically when the proposal density is well chosen. However, the behavior of MCMC algorithms is highly influenced by the choice of network statistics. In some cases, peculiar convergence properties have been observed, where the sequence of generated networks transitions rapidly between highly dissimilar states with occasional long periods of minor variation. Previous studies that have applied MCMC algorithms for social network models have faced challenges in attaining convergence to realistic distributions. As a consequence, the generated networks may exhibit unusual properties such as being complete (containing all possible edges), empty (containing no edges), or displaying extreme characteristics. It is crucial to recognize that these behaviors are inherent to the models being used and not solely attributable to the MCMC algorithms employed for simulation. These issues have been reported in the work of Hunter et al. (2008b), highlighting the complexities involved in capturing the true distribution of social networks through MCMC-based approaches.

5.7 Goodness-of-Fit for Dyadic Dependence Models

Assessing the goodness-of-fit for Exponential Random Graph Models (ERGMs) is crucial to determine how well the models capture the properties of observed networks. However, traditional methods like AIC (Akaike Information Criterion) or BIC (Bayesian Information Criterion) are not appropriate for network data because they assume that the samples are independent and identically distributed, which is not the case for network observations. An alternative approach, proposed by Hunter et al. (2008b), evaluates the goodness-of-fit for ERGMs through visual plots. The underlying concept is to compare the network statistics observed in the original data with the statistics obtained from networks simulated using the fitted ERGM. This comparison allows for assessing whether the simulated structural statistics closely match those observed, indicating a good fit of the ERGM to the data. The first step in assessing the goodness-of-fit is to estimate the ERGM using the observed network data. This involves fitting the ERGM to the observed network, specifying the appropriate network statistics and their corresponding model terms. The estimation process aims to find the parameter values that best capture the observed network's characteristics. Once the ERGM is estimated, the next step is to simulate multiple networks from the fitted model using the estimated parameters. These simulated networks are generated to mimic the structure and properties of the observed network as captured by the ERGM. The number of simulated networks to generate can be adjusted based on the desired level of precision. For both the observed and simulated networks, various network statistics are calculated. These statistics quantify different aspects of the network's structure, such as the number of edges, clustering coefficient, degree distribution, or any other relevant measures. The goal is to compute these statistics for both the observed and simulated networks consistently to facilitate meaningful comparisons. To visually compare the observed and simulated network statistics, plots are constructed. Each plot represents a specific network statistic, with the observed structure statistics plotted against the corresponding simulated structure statistics. This graphical representation allows for an intuitive assessment of the similarity or dissimilarity between the observed and simulated statistics. By analyzing the plots, one can evaluate the degree of similarity between the observed and simulated statistics. A good fit of the ERGM would result in the observed and simulated statistics exhibiting a close alignment or following a similar pattern in the plots. On the other hand, substantial deviations or discrepancies between the observed and simulated statistics may indicate a poor fit of the ERGM to the data. Through this iterative process of estimating the ERGM, simulating networks, calculating network statistics, and comparing observed versus simulated statistics in plots, we are able to gain insights into how well the ERGM captures the key characteristics of the observed network. This evaluation aids in validating the model's adequacy and identifying areas where the model may need improvement to better represent the observed network structure.

5.7.1 R Code and Output

```
# Install and load the ERGM and network packages
> install.packages("ergm")
> install.packages("network")
> library(ergm)
> library(network)

# Set the number of nodes and covariates
> num_nodes <- 60
> num_edges <- 800
> num_covariates <- 2

# Generate random covariate values
> covariate1 <- rnorm(num_nodes)
> covariate2 <- rbinom(num_nodes, 1, 0.5)

# Generate a random sociomatrix (adjacency matrix)
> sociomatrix <- matrix(sample(c(0, 1), num_nodes^2, replace = TRUE, prob = c(0.8, 0.2)), nrow = num_nodes)

# Convert sociomatrix to a network object
> network <- as.network(sociomatrix, directed = FALSE)

# Add node covariates to the network
> network %v% "Covariate1" <- covariate1
> network %v% "Covariate2" <- covariate2

# Define the ERGM formula with
> formula <- network ~ edges + triangles + nodematch("Covariate1") + nodematch("Covariate2")

# Fit the ERGM
> model <- ergm(formula, data = network)

# Print the model summary
> summary(model)

Output

  Call:
  ergm(formula = formula, data = network)

  Monte Carlo Maximum Likelihood Results:

                       Estimate Std. Error MCMC % z value Pr(>|z|)
  edges                -0.55173    0.24433      0  -2.258   0.0239 *
  triangle             -0.01810    0.03352      0  -0.540   0.5892
  nodematch.Covariate1     -Inf    0.00000      0    -Inf   <1e-04 ***
  nodematch.Covariate2  0.09752    0.10602      0   0.920   0.3576
  ---
  Signif. codes:  0 '***' 0.001 '**' 0.01 '*' 0.05 '.' 0.1 ' ' 1

   Warning: The following terms have infinite coefficient estimates:
    nodematch.Covariate1
```

In the RStudio implementation just shown, we have the results but at the end we have a warning. The warning message indicates that the term *nodematch.Covariate1* in our ERGM model has an infinite coefficient estimate. This means that the estimated coefficient for this term is either extremely large (positive or negative) or undefined. In ERGMs, the *nodematch* term is used to capture the tendency of nodes with similar attribute values to form ties with each other. In our case, the *nodematch. Covariate1* term specifically models the tendency for nodes with similar values of *Covariate1* to be connected. When the coefficient estimate for a term in an ERGM is infinite, it typically indicates perfect or near-perfect separation in the data. This means that the attribute being considered (in this case, *Covariate1*) perfectly distinguishes between the presence and absence of ties in the network. As a result, the coefficient estimate cannot be reliably estimated and is set to infinity. There are a few reasons why you may be observing this behavior:

- **Perfect separation**: The values of *Covariate1* in your network may perfectly separate the presence and absence of ties. For example, if all nodes with a specific value of *Covariate1* are connected, while nodes with other values are not, the coefficient estimate for *nodematch.Covariate1* would be infinite.
- **Small sample size**: If our network has a small number of nodes or a small number of ties, it can increase the likelihood of encountering perfect or near-perfect separation. With limited data, it becomes more challenging to estimate the coefficient accurately.
- **Collinearity**: *Covariate1* may be highly correlated with other variables or terms in the model, leading to instability in the coefficient estimation. Collinearity can make it difficult for the model to differentiate the effects of individual covariates accurately.

The warning is generated in this example because nodematch is commonly employed to model the homophily effect when the variable is discrete. It does not instantly apply to continuous covariates like *Covariate1*. As a result, we have to employ a function designed for continuous data, *absdiff* (absolute difference). It is important to select the appropriate term that aligns with the nature of your data. Below is the right function in RStudio.

5.7.2 R Code and Output

```
# Define the ERGM formula with the right function
> formula1 <- network ~ edges + triangles
> formula2 <- network ~ edges + triangles + absdiff("Covariate1") + nodematch("Covariate2")


# Fit the ERGM
> model1 <- ergm(formula1, data = network)
> model2<- ergm(formula2, data = network)

# Print the model summary
> summary(model1)
> summary(model2)


Output

ergm(formula = formula, data = network)

Monte Carlo Maximum Likelihood Results:

                       Estimate Std. Error MCMC % z value Pr(>|z|)
edges                  -0.57561    0.23869      0  -2.412   0.0159 *
triangle               -0.01859    0.03165      0  -0.587   0.5571
absdiff.Covariate1      0.03485    0.05961      0   0.585   0.5588
nodematch.Covariate2    0.07289    0.10136      0   0.719   0.4720
---
Signif. codes:  0 '***' 0.001 '**' 0.01 '*' 0.05 '.' 0.1 ' ' 1

     Null Deviance: 2454  on 1770  degrees of freedom
 Residual Deviance: 2284  on 1766  degrees of freedom

AIC: 2292  BIC: 2314  (Smaller is better. MC Std. Err. = 0.1156)

# Using the screenreg() function from the texreg package, we compare the results of the two estimations
> library(texreg)
> screenreg(list(model1, model2))
```

In addition to the terms mentioned earlier, namely *nodematch* and *absdiff*, there are several other ERGM terms available to capture different aspects of network structure and attribute associations. Here are some additional terms you can consider:

- *nodematch()*: This term captures homophily or the tendency for nodes with similar attribute values to be connected. It is suitable for modeling discrete attributes.
- *absdiff()*: This term is used for quantitative nodal attributes. It calculates the absolute difference between attribute values of connected nodes and includes it as a network statistic in the model.

$$\text{abs}\left(\text{attr}\left[\text{i}\right]-\text{attr}\left[\text{j}\right]\right)^{\text{pow}} \tag{5.51}$$

Here, *attr* represents the nodal attribute of interest, *i* and *j* represent the nodes connected by an edge, and *pow* is the power to which the absolute difference is raised. This term calculates the absolute difference between the attribute values of connected nodes and raises it to the specified power.

- *absdiffcat()*: Similar to *absdiff()*, this term is used for categorical nodal attributes. It captures the differences between attribute categories of connected nodes.

- *nodecov()*: This term incorporates quantitative nodal attributes directly as covariates in the model. It allows you to model the effect of specific nodal attributes on the edge probability.
- *nodefactor()*: This term is used for discrete attributes. It models the effect of a variable on the edge probability by treating it as a factor. Each level of the factor is assigned a separate coefficient.
- *nodemix()*: This term is used to determine the likelihood of a tie forming between nodes that have the same value of the categorical variable and those that have a different value of the same variable.
- *netcov()*: This term allows you to include a matrix of edge covariates in the model. Edge covariates are attributes that are specific to each edge in the network, such as the strength of the tie or the geographical distance between nodes.

These are just a few examples of the available ERGM terms. Each term serves a specific purpose and can be used to capture different types of network dynamics and attribute associations. You can refer to the *ergm-terms* ("?ergm-terms") documentation in RStudio for a comprehensive list of available terms and their usage.

5.8 Marginal Effects in ERGM for Binary Data

In Exponential Random Graph Models (ERGMs), the marginal effects of a covariate represent the effect of that covariate on the probability of a tie between two nodes in a network, holding all other variables constant. Marginal effects are important in ERGMs because they allow the quantification of the effect of a specific covariate on the network while controlling for the effects of all other covariates in the model. The marginal effect of a covariate in an ERGM is the change in the probability of a tie between two nodes in the network when the value of that covariate changes by one unit, while holding all other covariates constant. For example, if the effect of gender on the probability of a tie between two individuals in a social network is of interest, the marginal effect of gender would tell us how much the probability of a tie changes when moving from a network where all individuals are male to a network where all individuals are female, while holding other covariates constant.

Marginal effects can be estimated using various methods, such as simulation-based methods or closed-form solutions, depending on the complexity of the model and the type of covariate being considered. The interpretation of marginal effects can be complex in ERGMs, especially when there are multiple covariates in the model, and care must be taken to ensure that the effects are correctly interpreted and reported.

5.8.1 R Code and Output

```
# Define the ERGM formula
> formula <- network ~ edges +triangles+ nodecov("Covariate1")

# Fit the ERGM
> model <- ergm(formula, data = network)

# Print the model summary
> summary(model)

# Function to compute marginal effects at means
> install.packages("ergMargins")
> library(ergMargins)

# Main effect
> effect<-ergm.MEM(model,"triangle")
```

5.9 ERGM for Count Data

The ERGM for count data, as described by Krivitsky (2012, 2023), extends the basic ERGM framework to account for the count of edges in the network. ERGMs are often used to model binary network data, where edges represent the presence or absence of ties between nodes. However, in some cases, it may be more appropriate to consider count data, where the edges represent the number of connections between nodes.

Krivitsky (2012) generalized the ERGM to directly model valued networks, namely ties that represent counts. The ERGM for count data of *Y* has probability mass function (PMF):

$$\Pr_{\theta,\mathrm{h},\eta}\left(Y=y\right)=\frac{h\left(y\right)\exp\left(\eta\left(\theta\right)\cdot g\left(y\right)\right)}{\kappa_{h,\eta}\left(\theta\right)} \quad \text{with } y\in Y \tag{5.52}$$

where $Pr_{\theta,h,\eta}(Y = y)$ represents the probability mass function (PMF) of the random variable *Y*, parameterized by θ, in the exponential family form. It gives the probability of observing the value *y* of *Y*. *Y* is defined as some subset of *SY*, which is the total number of possible ways to assign to each link a value, for valued edges this can be $S = \{0,1,2...s\}$. In this case, the sample space is a set of mappings that are assigned to each dyad *i, j*. *h(y)* is a function that depends solely on the observed value *y*. It is called the base measure or base density and determines the scaling of the distribution. $exp(\eta(\theta).g(y))$ is an exponential function, where $\eta(\theta)$ is a vector of natural parameters (also known as canonical parameters) that depend on the parameter vector θ, and *g(y)* is a vector of sufficient statistics that depends on the observed value *y*. The product of $\eta(\theta)$ and *g(y)* ensures that the exponential function is properly scaled. Finally, $\kappa_{h,\eta}(\theta)$ normalizes the exponential family distribution. It ensures that the probabilities or densities integrate or sum up to 1 over the entire support of the random variable *Y*.

The denominator is a normalization constant defined as:

$$k_h(\theta) = \sum_{y \in \mathcal{Y}} h(y) \exp(\eta(\theta) \cdot g(y)) \tag{5.53}$$

with $h(y)$ that set up the baseline shape of the dyad distribution of response variable $y \rightarrow [0, \infty]$ (Krivitsky, 2012). It specifies the distribution relative to which the exponential form of the ERGM is defined. As an example, according to Krivitsky (2012), the function $h(y)$ has a discrete uniform distribution as the reference measure in a geometric reference ERGM for count data. In contrast, in a Poisson-reference ERGM, the function $h(y)$ is defined as the product of the inverse factorial of the count $y_{i,j}$ over all dyads (i, j) in the network. The reference measure "h" determines the baseline shape of the dyad distribution, which represents the probability distribution of the dyad values (counts) in the absence of any network structure or parameter effects. Different choices of "h" lead to different baseline shapes for the dyad distribution. Moreover, the reference measure "h" also constrains the parameter space of the ERGM. The parameters of the model are estimated by maximizing the log-likelihood function, which involves the normalization constant that depends on h. The choice of h affects the normalization constant and, consequently, the parameter space. By selecting different reference measures, such as geometric or Poisson reference, the baseline shape of the dyad distribution is altered. This, in turn, affects the types of network structures that can be modeled effectively and the range of parameter values that are feasible.

ERGMs for counts are inferentially similar to discrete exponential families and binary ERGMs. This is beneficial since it allows for the computation of standard errors and the analysis of deviation, both of which are often employed for statistical inference. However, when applying asymptotic tests, such as the Wald test, in ERGMs with complicated dependence structures, caution is advised because their validity may be called into doubt. To address the limitations of asymptotic tests, Krivitsky (2012) advises using a Monte Carlo test. A more robust p-value can be produced by fitting a nested model that excludes the statistic of interest and simulating its distribution. This method provides a practical way to assess the statistical significance of the observed statistic while mitigating the drawbacks of asymptotic tests. Asymptotic tests rely on large sample approximations and assumptions about the underlying distribution. However, these assumptions may not hold in complex network structures or when dealing with count data. Monte Carlo simulation can account for complex dependence structures and data patterns that may not conform to the assumptions of traditional tests. Furthermore, Krivitsky (2012, p. 1106) advises that using ERGMs with count data raises new inferential issues. The unbounded sample space of counts is one stumbling issue, which indicates that satisfying the limits of the parameter space may be problematic. The parameter space represents the valid range of parameter values that ensure the model remains consistent. By constraining the parameters within a specific space, the model captures the desired dependencies and relationships in the network data accurately. Because of this significant constraint, ERGMs with high-valued statistics may fail to meet regularity criteria, necessitating careful attention during the investigation.

Furthermore, computational issues arise when applying ERGMs to count data (for more information, see Krivitsky, 2012, p. 1107).

5.9.1 R Code and Output

```
# Required libraries
> library(network)
> require(ergm)
> library(ergm.count)

# Set the number of nodes and edges
> num_nodes <- 30
> num_edges <- 350

# Generate random discrete weights from 0 to 10
> weights <- sample(0:10, num_edges, replace = TRUE)

# Create an empty adjacency matrix
> adj_matrix <- matrix(0, nrow = num_nodes, ncol = num_nodes)

# Add edges with weights to the adjacency matrix
> for (i in 1:num_edges) {
  source_node <- sample(1:num_nodes, 1)
  target_node <- sample(setdiff(1:num_nodes, source_node), 1)
  adj_matrix[source_node, target_node] <- weights[i]
  adj_matrix[target_node, source_node] <- weights[i]  # If the graph is undirected
}

# Get the weighted adjacency matrix
> weighted_matrix <- adj_matrix

# Network with weighted edges. The weight is determined by the names.eval option
> net<-as.matrix(weighted_matrix)
> valuedNet<-as.network(net,directed= FALSE, ignore.eval=FALSE, names.eval='lag')

# Generate random covariate values
> covariate1 <- rnorm(num_nodes)
> covariate2 <- rbinom(num_nodes, 1, 0.5)

# Add node covariates to the network
> valuedNet %v% "Covariate1" <- covariate1
> valuedNet %v% "Covariate2" <- covariate2

# Fit the ERGM.COUNT
> model1 <- ergm(valuedNet~sum+ transitiveweights + absdiff("Covariate1") + nodematch("Covariate2") +
nodecov("Covariate1"), response="lag", reference=~Binomial(3))
> summary(model1,digits = max(1))

Output

 Monte Carlo Maximum Likelihood Results:

                              Estimate Std. Error MCMC % z value Pr(>|z|)
 sum                          -13.48245   41.86001      0  -0.322  0.74739
 transitiveweights.min.max.min 11.17609   41.85993      0   0.267  0.78948
 absdiff.sum.Covariate1         0.31280    0.11637      0   2.688  0.00719 **
 nodematch.sum.Covariate2      -0.01947    0.20114      0  -0.097  0.92287
 nodecov.sum.Covariate1         0.09688    0.07957      0   1.217  0.22343
 ---
 Signif. codes:  0 '***' 0.001 '**' 0.01 '*' 0.05 '.' 0.1 ' ' 1

      Null Deviance:    0.0  on 435  degrees of freedom
  Residual Deviance: -526.4  on 430  degrees of freedom

# MCMC diagnostic test
> mcmc.diagnostics(model1)
```

5.10 Degeneracy Issue

The use of Exponential Family Random Graph Model (ERGM) has been associated with a set of problems commonly known as "degeneracy" (Handcock et al., 2003; Rinaldo et al., 2009; Schweinberger, 2011). Degeneracy in ERGMs refers to a set of issues and limitations associated with modeling complex network structures. It arises from the inherent difficulty of capturing the full complexity of network data and the interdependencies among network ties. One of the key manifestations of degeneracy is the presence of multiple network configurations with similar or equal probabilities under the model. This means that different network structures can have the same likelihood, making it challenging to estimate the true underlying parameters accurately. Degeneracy often arises due to the high dimensionality and complexity of network data. As the number of nodes and possible network configurations increases, the space of possible networks becomes huge, and estimating the model parameters becomes more challenging. The estimation process may become trapped in local optima, leading to biased or inefficient parameter estimates. Several factors contribute to degeneracy in ERGMs. One factor is the presence of highly interconnected structures, such as triangles or stars, which exhibit strong positive dependencies. These structures can result in a concentration of probability on certain network configurations, leading to bimodal or multimodal distributions. As a result, ERGMs may assign high probabilities to both sparse and dense networks, making it difficult to capture the true network structure accurately. Another factor that contributes to degeneracy is the limited availability of information in the observed network data. Networks are typically sparse, meaning that only a small fraction of possible ties are present in the observed data. This sparsity poses challenges for estimating the parameters accurately, as the model has limited information to infer the underlying network structure. Additionally, degeneracy can also arise due to model misspecification or the presence of unobserved heterogeneity. If the chosen model inadequately captures the underlying network processes or fails to account for important features, it can result in biased or misleading parameter estimates. Specifically, Rinaldo, Fienberg, and Zhou identified three interrelated singularities:

- *Bimodal distribution*: In certain parameter configurations, including the maximum likelihood estimate (MLE), the induced distribution may have a small number of networks with significant probabilities, leading to a bimodal distribution. These networks often differ greatly from each other, such as a sparser-than-observed graph and a complete graph, resulting in a bimodal distribution.
- *Difficulty in finding the MLE*: The MLE can be challenging to find using available Markov chain Monte Carlo (MCMC) methods.

- *Low probability of the observed network*: The probability of the observed network under the MLE is relatively low, indicating that the observed network falls between the modes of the distribution.

These singularities arise due to the model inducing strong positive dependence among dyad values. For example, Snijders et al. (2006) demonstrate that ERGM terms with positive coefficients on triangle and k-star (with $k \geq 2$) counts lead to an "avalanche" effect. Adding a tie to the network increases the conditional odds of multiple other ties without decreasing the odds of any tie, resulting in a high probability of a complete graph. Adjusting other parameters, such as density, to match the sparsity level of the observed graph only exacerbates the bimodal distribution (first singularity).

While an infinite sample space (for instance, in the ERGM for count data) makes the emergence of a maximal graph unlikely, excessive dependence can still result in a bimodal distribution at the MLE, even if neither mode is degenerate in a probabilistic sense. This situation can lead to the observed network falling between the modes (third singularity) and can cause difficulties in estimation (second singularity) due to the nature of the estimation algorithms.

To address these issues, Krivitsky (2012) proposes the construction of statistics that limit or counterweight dependence to prevent the "avalanche" effect. Examples include modeling transitivity and employing centering in within-actor covariance statistics.

Addressing degeneracy in ERGMs is an active area of research. Various approaches have been proposed to mitigate degeneracy and improve the performance of models. These include incorporating additional network statistics or constraints to guide the estimation process, developing more efficient sampling algorithms, and using model selection techniques to identify the most appropriate model specification.

5.11 Temporal Exponential Random Graph Model (TERGM)

The Temporal Exponential Random Graph Model (TERGM) is an extension of the exponential random graph model that incorporates temporal dependencies and allows for the modeling of network evolution over time. TERGM is a discrete-time dynamic network model that represents dependencies in cross-sectional graphs and dependencies between graphs over time (Zinilli et al., 2024a). It postulates an exponential family model for the transition probability from a network at time t to a network at time $t + 1$. It is a statistical model for evolving networks that simplifies the calculation of probability by making a Markov Chain assumption on the network from one time step to the next. The probability of observing a network at time $t + 1$ given the network at time t is calculated using an exponential family model, which involves calculating the log-odds of each possible edge in the network. These log-odds are then exponentiated and normalized to obtain probabilities for each possible edge configuration. A network at time t, represented by the matrix Y^t, is assumed to be independent of all previous networks $(Y^1, Y^2, \ldots, Y^{t-2})$ given the network at the previous time step (Y^{t-1}). This assumption can be written as:

$$P\left(Y^2, Y^3, \ldots, Y^t, \middle| Y^1\right) = P\left(Y^t \middle| Y^{t-1}\right) \cdot P\left(Y^{t-1} \middle| Y^{t-2}\right) \cdot \cdots P\left(Y^2 \middle| Y^1\right) \tag{5.54}$$

Here, $P(Y^t \mid Y^{t-1})$ represents the one-step transition probability from the network Y^{t-1} to Y^t, and it is defined using a set of parameters (θ) and a sufficient statistic function $g(y)$. The formula for this transition probability is given by:

$$\mathrm{P}\left(\mathrm{Y}^{t} = \mathrm{y}^{t} \middle|, \mathrm{Y}^{t-1} = \mathrm{y}^{t-1} |, \theta\right) = \frac{\exp\left(\theta^{\mathrm{T}} \cdot g\left(\mathrm{y}^{t}, \mathrm{y}^{t-1}\right)\right)}{\sum_{\mathrm{y} \in \mathrm{Y}} \exp\left(\theta^{\mathrm{T}} \cdot g\left(\mathrm{y}^{t-1}\right)\right)} \tag{5.55}$$

In the above Eq. (5.55), θ is a q-dimensional vector of parameters, $g(y^t, y^{t-1})$ is the sufficient statistic function that captures the dependence between the networks at time t and $t-1$, and the summation factor $\Sigma_{y \in Y}$ensures that the equation represents a valid probability distribution. The choice of the sufficient statistic function $g(y^t, y^{t-1})$ depends on the specific network characteristics that are considered relevant for the transition between networks. It can be any valid network statistics evaluated on y^t and y^{t-1}, particularly those that depend on y^{t-1}. The selection of $g(y)$ is crucial in specifying the model and capturing the desired network dynamics. Hanneke et al. (2010) relied on a choice of $g(y)$ that has the attribute of conditional dyadic independence, namely:

$$P\left(Y^t = y^t \middle| Y^{t-1} = y^{t-1}, \theta\right) = \prod_{(i,j \in \Upsilon)} P\left(Y^t_{i,j} = y^t_{i,j} \middle| Y^{t-1} = y^{t-1}, \theta\right) \tag{5.56}$$

In this equation, $P(Y^t = y^t | Y^{t-1} = y^{t-1}, \theta)$ represents the conditional probability of Y^t being y^t given that Y^{t-1} is y^{t-1} and parameterized by θ. The product $\prod_{(i,j \in \Upsilon)}$ denotes taking the product over all indices (i,j) in the set Υ (the set of possible edges between them). $P\left(Y^t_{i,j} = y^t_{i,j} \middle| Y^{t-1} = y^{t-1}, \theta\right)$ represents the conditional probability of $Y_{(i,j)}$ being $y_{(i,j)}$ given that Y^{t-1} is y^{t-1} and parameterized by θ.

To estimate TERGM, two commonly used methods are Maximum Pseudolikelihood with Bootstrap Confidence Intervals (MPLE) and Markov Chain Monte Carlo Maximum Likelihood Estimation (MCMC-MLE). MPLE involves maximizing the pseudolikelihood function, which approximates the true likelihood by assuming conditional independence of events given the network's current state. To obtain valid confidence intervals, MPLE employs bootstrap resampling, generating multiple bootstrap samples from the original dataset. Each bootstrap sample is fitted to the TERGM model, resulting in a distribution of parameter estimates from which confidence intervals can be constructed. As noted in paragraph 5.5, MCMC-MLE calculates the real probability of the observed data directly without making the assumption of conditional independence. It generates samples from the posterior distribution of the model parameters using Markov Chain Monte Carlo techniques. The MCMC technique converges to the posterior distribution after a sufficient number of iterations, producing accurate and consistent parameter estimations. It is essential for the model fitting to converge in order to get accurate parameter estimations. Converged models show that the estimating process has stabilized, enabling useful investigation of the evolution of social network dynamics.

For calculating TERGM using the Maximum Pseudolikelihood with Bootstrap Confidence Intervals (MPLE) and MCMC-MLE approaches, R software offers specialized packages and functions. The *btergm* function for the MPLE approach enables TERGM estimation using bootstrap confidence intervals. The *mtergm* function is employed for the MCMC-MLE technique.

The interpretation of parameters in these models requires careful consideration due to their complex nature. To provide further clarity on this concept, let's examine a few specific cases. Consider a statistical model that includes the edge count with a higher coefficient. This means that for any given previous network state (y^{t-1}), the model will assign a higher probability to networks with more edges. This effect operates in two simultaneous ways. To gain a deeper understanding of the influence of this coefficient, we can analyze its effects separately on both edge formation and edge preservation. First, an increase in the coefficient would enhance the weight given to networks where more edges are formed on previously empty connections. In other words, it increases the likelihood of new connections being established between individuals who were not previously connected. This aspect of the coefficient relates to the incidence or rate at which new edges are formed in the network. Second, a higher coefficient also increases the weight given to networks where more existing edges are preserved or fewer edges are dissolved. This implies that connections that already exist in the network are more likely to persist over time. Thus, the coefficient affects the duration or how long the connections tend to last once they are formed. In summary, a higher coefficient on the edge count statistic simultaneously increases the likelihood of new connections being formed (incidence) and existing connections being preserved (duration) within the network.

An illustration of a statistic called "stability", which controls the rate of development inside a network, was provided by Hanneke et al. (2010). This statistic, which measures the number of edge variables that do not change across time steps, is normalized by dividing it by the maximum number of edges that any one actor may have. This measure of stability is as follows:

$$\Gamma\left(y_{ij}^{t},y_{ij}^{t-1}\right)=\frac{1}{n-1}\sum_{(i,j)\in Y}\left(y_{ij}^{t},y_{ij}^{t-1}+\left(1-y_{ij}^{t}\right)\left(1-y_{ij}^{t-1}\right)\right) \tag{5.57}$$

Changing the coefficient of the stability metric will have a significant impact on the evolution of the network. Increasing the coefficient introduces a decelerating effect on the network, resulting in a slow and deliberate progression of changes over time. A smaller coefficient, on the other hand, functions as an accelerator, driving the network toward quicker and more rapid alterations. Manipulating the coefficient has implications for both the frequency and duration of links within the network. When the coefficient is higher, the occurrence of new links decreases, leading to a lower rate of their formation. At the same time, existing edges are less likely to dissolve, resulting in longer lifespans for these connections. Conversely, a lower coefficient encourages a higher occurrence of new edges but may also increase the likelihood of edge dissolution, leading to shorter lifespans for the connections. While TERGM provides valuable insights into network dynamics, it does not explicitly disentangle the effects of edge formation and dissolution (Zinilli et al., 2024a). This is where the Separable Temporal Exponential Random Graph Model (STERGM) comes into play. The fact that the stability term results in an uncertain interpretation of its related parameter might serve as motivation for this approach. STERGM is able to disentangle whether the network can be considered "stable" due to the absence of link dissolution or because of the absence of link formation.

5.11.1 R Code and Output

```
# Load the required libraries
> library(tergm)
> library(network)

# Make this example reproducible
> set.seed(7)

# Set the number of nodes, edges and covariates
> num_nodes <- 15
> num_edges <- 105

# Generate random covariate values
> num_covariates <- 4
> covariates <- list()
> for (i in 1:num_covariates) {
  covariates[[i]] <- rnorm(num_nodes)
}
> covariate <- rbinom(num_nodes, 1, 0.4)

# Generate random sociomatrices (adjacency matrices) and convert them to network objects
> networks <- list()
> for (i in 1:num_covariates) {
  sociomatrix <- matrix(sample(c(0, 1), num_nodes^2, replace = TRUE, prob = c(0.8, 0.2 + i * 0.5)), nrow = num_nodes)
  networks[[i]] <- as.network(sociomatrix, directed = FALSE)
}

# Set vertex attributes for covariates
> for (i in 1:num_covariates) {
  covariate_name <- paste0("covariate", i)
  set.vertex.attribute(networks[[i]], covariate_name, covariates[[i]])
  set.vertex.attribute(networks[[i]], "covariate", covariate)
}

# Create a list of covariate lists
> list_cov <- list()
> for (i in 1:num_covariates) {
  list_cov[[i]] <- covariates[[i]]
}

# Assign covariate lists and covariate attribute to each network
> for (i in 1:num_covariates) {
  networks[[i]] <- set.vertex.attribute(networks[[i]], "list_cov", list_cov)
  networks[[i]] <- set.vertex.attribute(networks[[i]], "covariate", covariate)
}

# Compute the size of each network
> network_sizes <- sapply(networks, network.size)

# Fit the TERGM model to the networks
> model1 <- mtergm(networks ~ edges + gwesp(0.1, fixed = TRUE) + memory(type = "stability", lag = 1) +
timecov("covariate", transform = function(t) t) + absdiff("list_cov")+nodemix("covariate") , control = control.ergm(parallel
= 2, parallel.type = "PSOCK"), R = 50)

# Print the model summary
> summary(model1)

Output

    Time steps: 3

    Monte Carlo MLE Results:
                        Estimate  Std. Error  t value Pr(>|t|)
    edges             -6.5362e+05 3.8807e+04 -16.8427  < 2e-16 ***
    gwesp.fixed.0.1    5.9142e+05 3.5114e+04  16.8428  < 2e-16 ***
    edgecov.memory     2.2403e-01 2.4382e-01   0.9188  0.35853
    edgecov.timecov1   5.8826e-01 3.3495e-01   1.7562  0.07954 .
    absdiff.list_cov  -3.3546e-01 2.2986e-01  -1.4594  0.14496
    mix.covariate.0.1 -7.2575e-01 5.6283e-01  -1.2895  0.19771
    mix.covariate.1.1 -7.2091e-01 6.7314e-01  -1.0710  0.28460
    ---
    Signif. codes:  0 '***' 0.001 '**' 0.01 '*' 0.05 '.' 0.1 ' ' 1
```

```
# Perform goodness-of-fit assessment
> target_network <- networks[[num_covariates]]
> gof.model <- btergm::gof(model1, nsim = 100, target = target_network, formula = networks ~ edges + gwesp(0.1, fixed =
TRUE) + memory(type = "stability", lag = 1) + timecov("covariate", transform = function(t) t) +
absdiff("list_cov")+nodemix("covariate") , coef = coef(model), statistics = c(esp, dsp, deg, rocpr))

# Plot the goodness-of-fit assessment results
> plot(gof.model, roc.rgraph = TRUE, pr.rgraph = TRUE)
```

Output

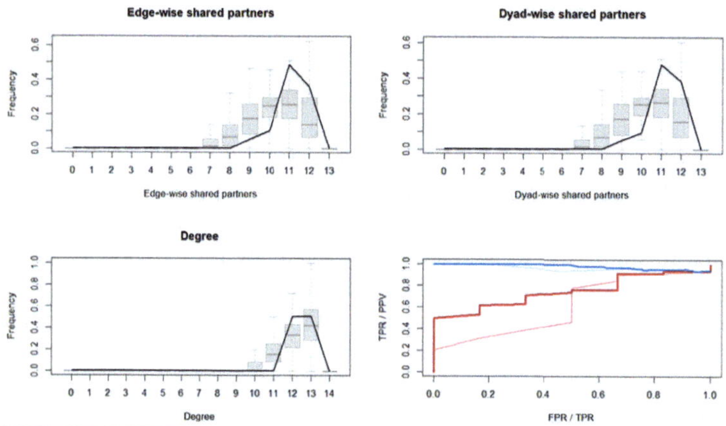

The variable *"gwesp"* has a positive estimate, suggesting that the presence of triads in the network has a positive impact. The positive estimate for *"edgecov.timecov1"* indicates that the time-related covariate has a positive impact on the network. This implies that changes in the time-related factor are associated with changes in the network structure. Regarding the other variables, their estimates are not statistically significant. This implies that these variables do not have a significant influence on the network dynamics.

The GOF model calculated the goodness-of-fit statistics for the given model. The *"nsim"* parameter specifies the number of simulations to be performed. The *"target_network"* is the observed network data that is being compared against the model. The formula specified in the *"formula"* option represents the model formula. The *"statistics"* parameter indicates the specific statistics that will be calculated for the goodness-of-fit evaluation, including esp. (edge-wise shared partner), *dsp* (dyad-wise shared partner), *deg* (degree distribution), and *rocpr* (receiver operating characteristic precision-recall curve—ROC curve).

In our model, the fit of these statistics, particularly the ROC curve, is not consistently satisfactory. This suggests that the model may not accurately capture the relationship between the variables or adequately predict the network structure. It indicates that there may be limitations or shortcomings in the model's ability to explain the observed data and its underlying patterns. Since the data used for analysis is not derived from actual observations, the model's performance and its ability to capture network dynamics are limited.

5.12 The Separable Temporal Exponential Random Graph Model (STERGM)

The separable temporal exponential random graph model offers a more nuanced approach by introducing the concept of separability, allowing researchers to examine the individual contributions of formation and dissolution processes to network evolution. By incorporating this level of granularity, STERGM enhances researchers' analytical capabilities and facilitates a deeper understanding of the mechanisms shaping dynamic networks. The Separable Temporal Exponential Random Graph model was first introduced by Krivitsky and Handcock (2014). In STERGM, two ERGMs are used simultaneously: one to capture the process of tie formation and the other to capture the process of tie dissolution. This approach recognizes that the mechanisms driving the formation and dissolution of ties in social networks are likely to be distinct. For instance, the link formation in ERGM incorporates parameters related to similarity, which capture the tendency for individuals with similar characteristics to form connections. On the other hand, these similarities parameters may be less critical when modeling tie dissolution. The features affecting the probability of relationship dissolution may not perfectly mirror those, influencing relationship formation. The application of STERGM allows for a more nuanced understanding of the dynamics within social networks, as it recognizes that tie formation and tie dissolution processes can have distinct underlying mechanisms. To achieve this separation, Krivitsky and Handcock (2014) define the formation network, denoted as Y^+, which combines the initial network Y_{t-1} with all newly added ties in time period *t*. Conversely, the dissolution network, denoted as Y^-, consists only of ties that are present in both time periods t and *t−1*. By considering the network in time period *t−1* alongside the formation and dissolution networks, it becomes possible to reconstruct the network in time period *t* as $Y_t = Y^- \cup (Y^+ \setminus Y_{t-1})$.

In order to model the separation of formation and dissolution, Krivitsky and Handcock (2014) introduce a joint parameter vector $\theta = (\theta^+, \theta^-)$ that encompasses the parameters for the formation and dissolution models, respectively. The separability assumption states that the parameter space of θ is the Cartesian product of the parameter spaces of θ^+ and θ^-, and that the formation (red rectangle) and dissolution (blue rectangle) processes are conditionally independent given the network in time period *t−1*:

$$P\theta\left(Y_t = y_t \middle| Y_{t-1} = y_{t-1}, x_t\right) = \boxed{P\theta^+\left(Y^+ = y^+ \middle| Y_{t-1} = y_{t-1}, x_t\right)} * \boxed{P\theta^-\left(Y^- = y^- \middle| Y_{t-1} = y_{t-1}, x_t\right)} \quad 5.58$$

Figure 5.2 shows the network state at time interval *t−1* on the left and the resultant network at time point *t* on the right to give a visual representation of this model.

The top network showcases the possible ties that could be formed (formation), while the bottom network represents the ties that could be dissolved (dissolution).

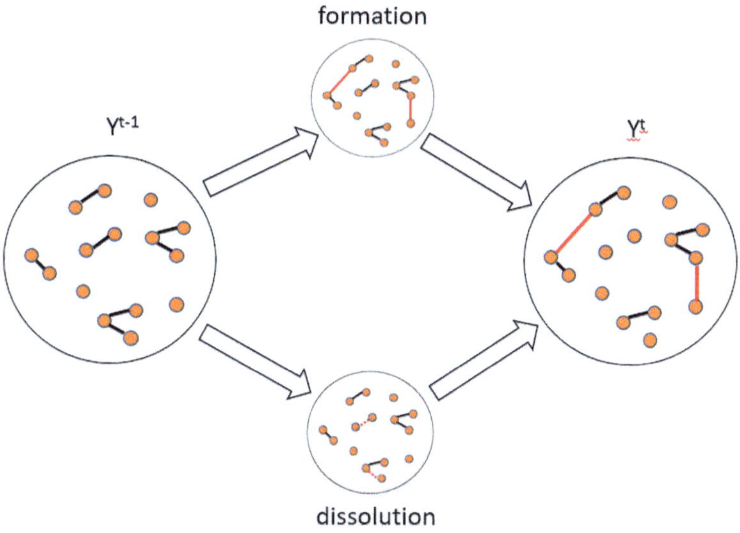

Fig. 5.2 Formation and dissolution representation in STERGM

The next Eq. (5.59) represents the conditional probability of the formation network, Y^+, given the network at time $t-1$ (Y_{t-1}), and the covariate vector x_t. This probability is computed using the formation model in the STERGM framework:

$$P\theta^+\left(Y^+ = y^+ \middle| Y_{t-1} = y_{t-1}, x_t\right) = \frac{\exp\left\{\left(\theta^+\right)^T g\left(y^+, y_{t-1}, x_t\right)\right\}}{\kappa\left(\theta^+, y_{t-1}, x_t\right)} \tag{5.59}$$

where $\theta+$ represents the parameter vector specific to the formation model and $g(y_+, y_{t-1}, x_t)$ is a vector of network statistics that captures the structural properties of the formation network. Finally, $\kappa(\theta^+, y_{t-1}, x_t)$ is the normalization constant ensuring that the probabilities sum to 1. Similarly, the dissolution model can be defined.

$$P\theta^-\left(Y^- = y^- \middle| Y_{t-1} = y_{t-1}, x_t\right) = \frac{\exp\left\{\left(\theta^-\right)^T g\left(y^-, y_{t-1}, x_t\right)\right\}}{\kappa\left(\theta^-, y_{t-1}, x_t\right)} \tag{5.60}$$

Therefore, in the dissolution model, we have the probability of observing the dissolution network y^-, given the previous network y_{t-1} and the covariate vector x_t, based on the parameters θ^- and the associated network statistics. It is worth noting that by incorporating the separable models into Eq. (5.58), we can express the probability of the network at time period t as:

$$P\theta^{+}\left(Y^{+}=y^{+}\middle|Y_{t-1}=y_{t-1},x_{t}\right)=\frac{\exp\left\{\left(\theta^{+}\right)^{T}g\left(y^{+},y_{t-1},x_{t}\right)\right\}}{\kappa\left(\theta^{+},y_{t-1},x_{t}\right)}^{*}\frac{\exp\left\{\left(\theta^{-}\right)^{T}g\left(y^{-},y_{t-1},x_{t}\right)\right\}}{\kappa\left(\theta^{-},y_{t-1},x_{t}\right)}$$
$$=\frac{\exp\left\{\left(\theta^{T}\right)g\left(y_{t},y_{t-1},x_{t}\right)\right\}}{\kappa\left(\theta,y_{t-1},x_{t}\right)} \tag{5.61}$$

where $\theta = (\theta^{+}, \theta^{-})^{T}$ is the joint parameter vector that combines the parameters of the formation and dissolution models. The vector of network statistics, $g(y_t, y_{t-1}, x_t)$, captures the relevant structural properties of the network at time t and the network history up to time $t-1$. Hence, through the incorporation of separable models, the probability of the network at time period t can be represented as an exponential function of the joint parameter vector θ and the pertinent network statistics. This probability is then divided by the normalization constant, which ensures proper scaling and validity. A positive coefficient in the dissolution model indicates that nodes or dyads with higher values for the associated statistic are actually less likely to dissolve. It is important to note that this convention is widely followed in the Rstudio, but if desired, the interpretation can be reversed simply by changing the signs of the parameters in the dissolution model. By adopting this framework, we gain the ability to model both the processes of link formation and dissolution, while effectively considering the unique parameters and structural characteristics associated with each process.

5.12.1 R Code and Output

```
# Load the required libraries
> library(tergm)
> library(network)

# Make this example reproducible
> set.seed(7)

# Set the number of nodes, edges, and covariates
> num_nodes <- 15
> num_edges <- 80

# Generate random covariate values
> num_covariates <- 4
> covariates <- list()
> for (i in 1:num_covariates) {
 covariates[[i]] <- rnorm(num_nodes)
}
> covariate <- rbinom(num_nodes, 1, 0.2)

# Generate random sociomatrices (adjacency matrices) and convert them to network objects
> networks <- list()
> for (i in 1:num_covariates) {
  sociomatrix <- matrix(sample(c(0, 1), num_nodes^2, replace = TRUE, prob = c(0.8, 0.8 + i * 0.5)), nrow = num_nodes)
  networks[[i]] <- as.network(sociomatrix, directed = FALSE)
}

# Set vertex attributes for covariates
> for (i in 1:num_covariates) {
  covariate_name <- paste0("covariate", i)
  set.vertex.attribute(networks[[i]], covariate_name, covariates[[i]])
  set.vertex.attribute(networks[[i]], "covariate", covariate)
}

# Create a list of covariate lists
> list_cov <- list()
> for (i in 1:num_covariates) {
  list_cov[[i]] <- covariates[[i]]
}

# Assign covariate lists and attributes to each network
> for (i in 1:num_covariates) {
  networks[[i]] <- set.vertex.attribute(networks[[i]], "list_cov", list_cov)
  networks[[i]] <- set.vertex.attribute(networks[[i]], "covariate", covariate)
}

# Compute the size of each network
> network_sizes <- sapply(networks, network.size)

# Fit the STERGM model to the networks
> model2 <- stergm(networks, formation = ~ edges + gwesp(0.1, fixed = TRUE) +nodecov("list_cov") +
nodematch("covariate") , dissolution = ~ edges + gwesp(0.1, fixed = TRUE)+ nodecov("list_cov")+ nodematch("covariate") ,
estimate = "CMLE")

# Print the model summary
> summary(model2)

Output

     Monte Carlo Conditional Maximum Likelihood Results:

                                Estimate Std. Error MCMC %    z value Pr(>|z|)
     Form~edges               -1.693e+09  1.005e+04    100 -1.683e+05   <1e-04 ***
     Form~gwesp.fixed.0.1      1.532e+09  9.098e+03    100  1.683e+05   <1e-04 ***
     Form~nodecov.list_cov    -3.929e-01  2.685e-01      0 -1.463e+00    0.143
     Form~nodematch.covariate -3.260e-01  1.240e+00      0 -2.630e-01    0.793
     Persist~edges            -2.370e+03  2.429e+02    100 -9.757e+00   <1e-04 ***
     Persist~gwesp.fixed.0.1   2.146e+03  2.198e+02    100  9.766e+00   <1e-04 ***
     Persist~nodecov.list_cov  1.040e-01  1.670e-01      0  6.230e-01    0.533
     Persist~nodematch.covariate 8.414e-01  5.583e-01    0  1.507e+00    0.132
     ---
     Signif. codes:  0 '***' 0.001 '**' 0.01 '*' 0.05 '.' 0.1 ' ' 1

          Null Deviance: 436.7  on 315  degrees of freedom
      Residual Deviance: 144.0  on 307  degrees of freedom

     AIC: 160  BIC: 190  (Smaller is better. MC Std. Err. = 0.4306)
```

The output is the result of a Monte Carlo Conditional Maximum Likelihood analysis, which includes estimates, standard errors, MCMC percentages, z-values, and p-values for several variables. The estimates represent the coefficients estimated for each variable in the example, while the standard errors indicate the variability or uncertainty associated with each estimate. The MCMC percentage indicates the proportion of Monte Carlo Markov Chain iterations used in the analysis, while z-values measure the number of standard deviations that an estimate is away from the null hypothesis. Lower p-values indicate statistical significance. The *Null Deviance* and *Residual Deviance* values provide information about the goodness-of-fit of the model. The *Null Deviance* represents the deviance when only the intercept is included in the model, while the *Residual Deviance* represents the deviance after including the predictors. Lower *Residual Deviance* indicates a better fit of the model to the data. The results of the example indicate that the presence of triads or closed triangles in the network significantly influences both the formation and dissolution of relationships. Specifically, the variable *gwesp* shows a positive estimate for both formation and dissolution, suggesting that the existence of these interconnected structures promotes the establishment and termination of relationships. This finding highlights the importance of triads or closed triangles in shaping network dynamics. However, the other variables, representing additional factors or characteristics of the nodes, do not show statistically significant estimates. This means that these exogenous variables do not play a significant role in the formation and dissolution processes. Overall, these findings emphasize the significance of network structure, specifically the presence of triads or closed triangles, in driving the formation and dissolution of relationships.

5.13 Generalized Exponential Random Graph Model (GERGM)

While ERGMs are primarily designed for binary networks, the Generalized Exponential Random Graph Model (GERGM) offers a more flexible framework to analyze networks with continuous or valued edges. As mentioned before, in traditional ERGMs, the focus is on modeling the presence or absence of links between nodes. However, many real-world networks have additional information associated with the links, such as strengths or weights. GERGM addresses this limitation by enabling the modeling of these continuous values in conjunction with the network structure. To explain this model, we will follow the approach presented by Wilson et al. (2017) and Desmarais and Cranmer (2012).

There are two specification steps in the Generalized Exponential Random Graph Models(GERGM). In the first step, specify a tractable joint distribution for a restricted valued network *X* that captures the relevant dependencies. The network *X* has the same vertices as the network of interest *Y*, but its edge values are continuous and bounded between zero and one ($X \in [0, 1]^m$). The variables *Y* and *X* represent different networks in the context of the GERGM. X is a restricted valued network with the same vertices as the network of interest Y. Its edge values are continuous and bounded between zero and one. X captures the dependencies of interest and is used to establish a tractable joint distribution in the modeling process. It acts as an intermediate step in the GERGM approach. Y represents the network of interest, which is the main focus of analysis and modeling. It shares the same vertices as X but may have different characteristics and interpretations for its edge values. The GERGM aims to model and analyze the dependencies and characteristics of Y based on the dependencies captured in X. As result, X is a transformed version of Y, with continuous and bounded edge values, used to specify the joint distribution capturing dependencies.

To account for external factors as well as endogenous effects, a set of network statistics called g are established. Then, by modifying the ERGM formula, the probability distribution for X is created. To consider the network's limited range, some adjustments are needed. The distribution for X can be formulated as follows:

$$f_x(X,\theta) = \frac{\exp\{\theta^T g(X)\}}{\int_{(0,1)^m} \exp\{\theta' g(Z)\} dZ^T} \tag{5.62}$$

here $g: [0, 1]^m \rightarrow R^p$ indicates the network statistics and represents joint characteristics of Y in the distribution of X (it follows an exponential function), $\theta \in Rp$ is the unknown parameter that we aim to estimate (θ^T is the transpose of θ). The term $g(Z)$ dZ represents the integration of the function g with respect to the variable Z. It is used to calculate the probability distribution for X by summing or integrating over all possible values of Zv. The convergent sum of the denominator is ensured by the fact that the links are continuous and bounded within the interval (0, 1). An important property of the distribution specified for X is that when there are no dependencies in the network, $f(X)$ represents independent uniform random variables. This means that if the network statistics are correctly specified and $g = 0$, then X has no dependencies, implying independence among the edges in Y.

In the second step, a transformation is applied on the restricted valued network X to align it with the characteristics of the network Y. The transformation refers to a process in which the restricted valued network X is modified to match the characteristics or properties of the network of interest Y. This alignment ensures that the transformed network retains relevant features and captures the dependencies of interest present in Y. This transformation D^{-1} to each edge of X is parameterized and ensures a one-to-one mapping and monotonic increase. Specifically, Y is denoted as:

$$Y_{ij} = D_{ij}^{-1}\left(X_{ij}, \lambda_{ij}\right) \tag{5.63}$$

Through this transformation, the aim is to ensure that the characteristics and properties captured by X, such as the edge values and their boundedness, are appropriately represented in Y. The parameter λ_{ij} allows for fine-tuning the mapping between the corresponding edges of X and Y, enabling a more accurate representation of the marginal features of Y_{ij}. With the transformation applied to the restricted network X, it is possible to derive the specification for the GERGM while preserving the fundamental structure and modeling capabilities of the ERGM. In this derived specification, the parameter vector d is now defined on the transformed network, capturing the dependencies and characteristics of the network of interest, Y. GERGM may be written as:

$$f_Y\left(Y,\theta,\Lambda\right)=\frac{\exp\left\{\theta^T g\left(Y,\Lambda\right)\right\}}{\int_{(0,1)^m}\exp\left\{\theta' g\left(Z\right)\right\}dZ^T}\prod_{ij} d\left(Y_{ij},\lambda_{ij}\right) \tag{5.64}$$

Here, the model parameters θ capture the network dependencies, and the transformation parameters Λ capture the marginal characteristics of Y. The term $d(Y_{ij}, \lambda_{ij})$ represents the density function associated with the transformation D. When choosing the density function d, it is crucial to consider the specific characteristics of the network of interest, Y. The density function d should be selected based on factors such as the support (range of possible values), moments (e.g., mean, variance, skewness, and kurtosis), and how Y depends on covariate information (Zinilli & De Marchi, 2020).

Estimating the parameters in the model is a complex task due to the challenging integral in the denominator of Eq. 5.64. To overcome this, the Markov chain Monte Carlo maximum likelihood estimation (MCMC-MLE) is developed. For a more in-depth understanding of GERGM estimation, please refer to the paper by Desmarais and Cranmer (2012). The paper provides further insights into the methodologies and techniques employed in estimating the parameters of GERGM and offers a comprehensive overview of the model's applications and implications.

In the following code, you will find an example of how to build a GERGM model in R using randomly generated data. We will apply the GERGM package developed by Denny et al. (2017), which provides functions for estimating and diagnosing generalized exponential random graph models. Please note that you can find the code and more detailed information about the GERGM package on the GitHub repository by Matthew Denny. The repository can be accessed at the following URL: https://github.com/matthewjdenny/GERGM

5.13.1 R Code and Output

```
# Load the required packages and libraries
> install.packages( pkgs = c("BH","RcppArmadillo","ggplot2","methods",
                  "stringr","igraph", "plyr", "parallel", "coda", "vegan", "scales",
                  "RcppParallel","slackr"), dependencies = TRUE)

# The "devtools" package provides functions that allow us to install R packages directly from online repositories like GitHub. Once you have installed the 'devtools' package in R, you can use it to install the 'GERGM' package from GitHub by running the following command. This command will connect to the GitHub repository, download the package source code, and install it in your R environment.
> install.packages("devtools")
> devtools::install_github("matthewjdenny/GERGM")
> library(GERGM)

# Generate a random network with 20 nodes
> set.seed(12345)
> net <- matrix(runif(400, 0, 1), 10, 10)
> colnames(net) <- rownames(net) <- LETTERS[1:10]

# Generate two exogenous covariates
> covariate1 <- rnorm(10, 10, 2)
> covariate2 <- rnorm(10, 5, 1)

# Create a data frame for the node-level covariates
> node_level_covariates <- data.frame(Covariate1 = covariate1, Covariate2 = covariate2)
> rownames(node_level_covariates) <- LETTERS[1:10]

# Convert the covariate2 vector into a matrix
> network_covariate <- matrix(covariate2, nrow = 10, ncol = 1)

# Specify the GERGM model formula
> formula <- net ~ edges(method = "regression") + ttriads +
  nodecov("covariate1") +
  netcov(network_covariate)

# Estimate the GERGM model
> gergm_model <- gergm(
  formula,
  covariate_data = node_level_covariates,
  number_of_networks_to_simulate = 1000,
  thin = 1/10,
  proposal_variance = 0.5, # Adjust the proposal variance
  MCMC_burnin = 200000, # Increase the burn-in period
  seed = 123,
  convergence_tolerance = 0.01,
  force_x_theta_update = 2
)

# Print the model
> print(gergm_model)

# Generate estimate plot
> Estimate_Plot(gergm_model)
```

Output

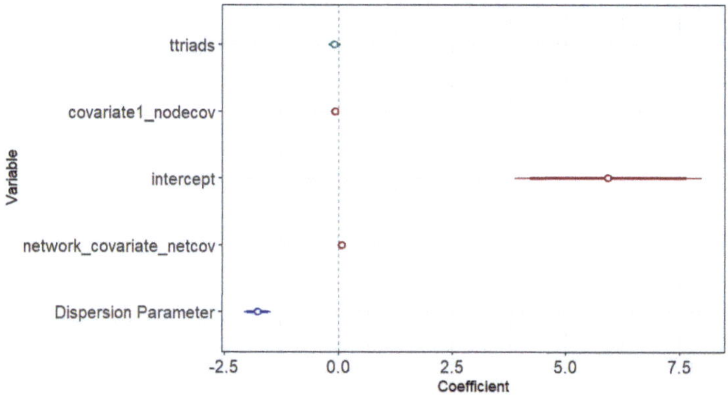

```
# Generate goodness-of-fit plot
> GOF(gergm_model)
```

Output

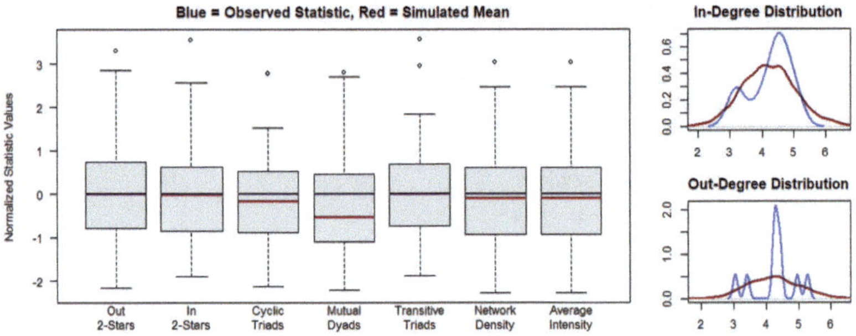


```
# Generate trace plot
> Trace_Plot(gergm_model)
```

Output

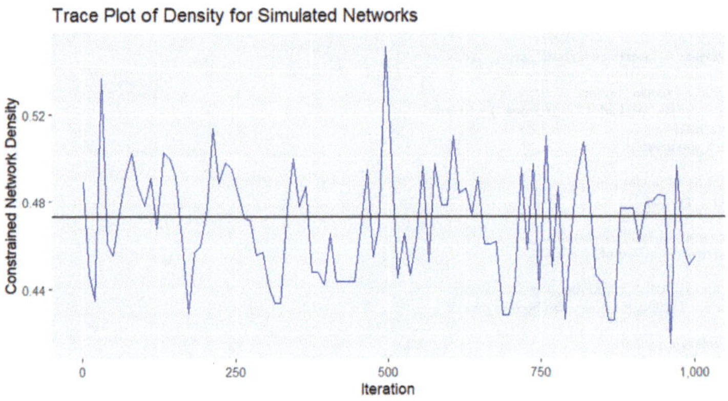


```
# Generate Hysteresis plots for the structural parameter estimates
> hysteresis_results <- hysteresis(gergm_model,
                   networks_to_simulate = 1500,
                   burnin = 600,
                   range = 2,
                   steps = 15,
                   simulation_method = "Metropolis",
                   proposal_variance = 0.2)
```

Output

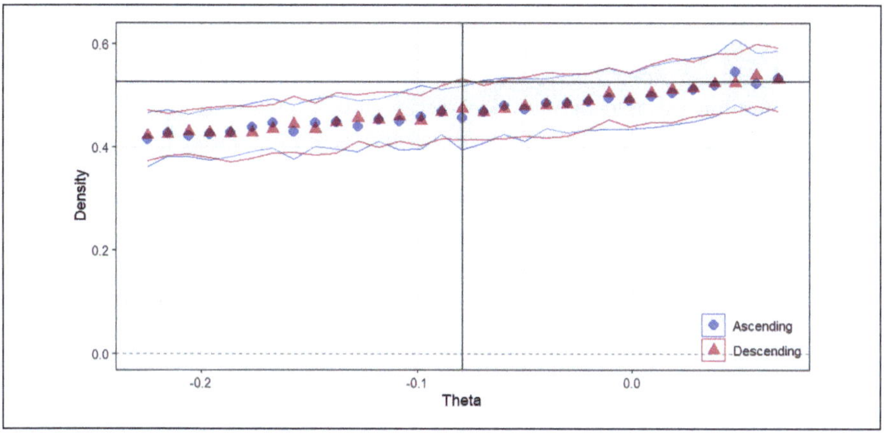

The *print* function provides an overview of the estimated parameters and statistics of the GERGM model. It displays the estimated values (est) and standard errors (se) for the different parameters in the model, such as *Theta*, *Lambda*, and *Weights*. Theta represents the effects of endogenous network statistics, Lambda represents the effects of node-level covariates and network covariates, and Weights represent the relative weights of the statistics used in the model.

The *Estimate_Plot* function generates a plot that visualizes the estimated parameter values from the GERGM model. This plot provides a graphical representation of the parameter estimates and their uncertainties, allowing for a visual assessment of the model's fit to the observed network data.

The *GOF* function generates a goodness-of-fit plot for the GERGM model. This plot compares the observed network statistics with the statistics simulated from the estimated model. It helps evaluate how well the model captures the patterns and features of the observed network. If the observed values fall within the range of the simulated distribution, it suggests that the model captures the patterns and features of the observed network. On the other hand, if the observed values deviate significantly from the simulated distribution, it indicates a lack of fit between the model and the data.

The *Trace_Plot* function generates a trace plot, which shows the MCMC samples of the parameter values over iterations. This plot helps assess the convergence of the MCMC algorithm and ensures that the samples explore the parameter space adequately. The x-axis represents the iteration number or the MCMC step, while the y-axis represents the sampled values of the parameter. Each point or line segment on the plot represents the value of the parameter at a specific iteration. By examining the trace plot, you can assess the convergence of the MCMC algorithm. Convergence refers to the property that the samples from the MCMC chain stabilize and accurately represent the posterior distribution of the parameters. In a trace plot, convergence is indicated by the samples fluctuating around a stable value or "mixing" well across the parameter space. If the trace plot shows erratic or unstable behavior, such as large fluctuations, long periods of flat or stagnant samples, or non-random patterns, it suggests issues with convergence. This could indicate insufficient exploration of the parameter space or other problems in the MCMC algorithm. On the other hand, a trace plot with smooth, well-mixed samples indicates good convergence. The samples should show reasonable variability and explore different regions of the parameter space, providing a reliable representation of the posterior distribution.

The *hysteresis()* function computes hysteresis plots for the structural parameter estimates of the GERGM model. A hysteresis plot is a useful tool for assessing the sensitivity of parameter estimates in a model. It helps to identify any signs of degeneracy, which refers to situations where multiple parameter combinations can produce similar model fits or likelihood values. When the hysteresis plot shows little or no signs of degeneracy, it indicates that the parameter estimates are stable and robust. This means that small changes in the parameter values do not result in significant changes in the model fit or likelihood values. It provides confidence that the estimated parameters are reliable and not highly influenced by minor variations (Snijders et al., 2006). When the points in the hysteresis plot follow a smooth upward trajectory (as in our example), it indicates that increasing the values of the parameters leads to a systematic and consistent change in the network structure. This suggests that the specification is not degenerate and is strongly supported by the evidence.

5.14 Stochastic Actor-Oriented Model

The Stochastic Actor-Oriented Model (SAOM) is a statistical framework used to analyze the social dynamics of network evolution. SAOM provides a way to model the mechanisms that underlie changes in social network structure over time, taking into account the individual characteristics and behaviors of actors within the network (Zinilli, 2016). The model is based on a probabilistic framework that incorporates both the structural properties of the network and the behavioral tendencies of actors, allowing for the examination of how these factors interact to shape network change. SAOM allows to incorporate additional features, such as dynamic effects, network autocorrelation, and the modeling of network dependencies, making it a powerful tool for the study of network dynamics (Snijders, 1996; Ripley et al., 2011). Built on Markovian processes, the model assumes that actors make choices based on the information they have about the relational structure within the network, without considering the network's past configurations as relevant. In practice, only information from time $t-1$ is utilized to understand the network configuration at time t, influencing actors' decisions and consequently shaping the future configuration. The initial observation time serves as the starting point, upon which the subsequent evolution depends, and thus is not explicitly modeled. Changes in the network occur step by step, that is, only one actor at a time makes only one change in their relational structure. These micro steps accumulate and give rise to significant changes. In the model, actors act to optimize the configuration of the network in relation to their utility and objectives (Huisman & Snijders 2003). The model assumes that all actors have a complete picture of the relational structure within the network. Based on this information, they evaluate their situation within the network of relationships and make changes to improve their position. Actors are independent of each other and act without coordination, only concerned with the short term and making no assumptions about the possible counter-moves of other actors; in short, they apply a myopic strategy. Assessments of the advantageous position within the network are a function of the actor's point of view (methodological individualism); in this sense, Snijders speaks of an actor-oriented model. The actor-oriented model is based on two statistical functions that describe the behavior of individual actors within a network. The first function, known as the time distribution function or *rate function*, specifies how often actors change their relationships within the network. This function may be uniform across all actors, or it may depend on specific attributes or positions within the network. The second function, called the *objective function*, determines the specific changes that an actor makes to their relationships. This function is made up of three parts: an evaluation function, which reflects an actor's satisfaction with different network configurations; an endowment function, which represents the benefits gained from making particular changes; and a random component that captures the effects of unmeasured factors.

In the next two sections, we will go into further depth about the *rate function* and *objective function*.

5.14.1 Main Assumptions

The Stochastic Actor-Oriented Model (SAOM) provides a theoretical basis for understanding how network ties evolve over time. It allows for the exploration of the interplay between individual actors' decisions and the dynamic nature of the network itself. The following assumptions form the foundation for analyzing and interpreting the results obtained from SAOMs in the study of networks in social systems (Amati et al., 2015; Snijders, 2017):

- *Ties are state*: This assumption recognizes that network ties in social contexts are not fleeting interactions but rather enduring relationships. By considering ties as states, we acknowledge that they can change but tend to endure over extended periods.
- *Distribution of the process*: The assumption of a continuous-time Markov chain implies that the network evolves over time in a probabilistic manner. Although we observe the network only at discrete time points, there exists an underlying latent process occurring between these observations. This continuous-time perspective allows for the consideration of the dependencies and temporal dynamics in the network evolution.
- *Opportunity to change*. At any given moment, one actor is selected probabilistically, and they have the opportunity to change one of their outgoing ties. This

assumption acknowledges that tie changes are solely driven by individual actors' choices and actions. It suggests that actors possess agency and can actively modify their behavior and connections within the network.

- *Absence of co-occurrence*. The assumption states that only one tie can change at a given moment, indicating that tie changes are not coordinated among multiple actors simultaneously. Instead, tie changes depend on each other sequentially, reflecting the evolving configuration of the entire network. This sequential nature implies that tie changes are influenced solely by the changing network structure and occur independently of each other.
- *Actor-oriented perspective*. The model takes an actor-oriented perspective, assuming that actors have control over their outgoing ties. This means that actors make decisions about changing their ties based on various factors, including their position in the network, personal attributes, and the characteristics of other actors. The assumption is that actors seek to maximize a utility function under the current network structure, aiming to achieve the most favorable outcomes by altering their connections. Additionally, it assumes that actors have complete knowledge about the network and other actors, enabling them to make informed decisions based on the current network configuration.

These assumptions collectively provide a framework for understanding the dynamic nature of social networks and the role of individual actors in shaping network evolution. They emphasize the enduring nature of social ties, the probabilistic nature of network changes, and the independence of individuals in making decisions regarding their social relationships. By considering these assumptions, we gain insights into the complex interplay between individual actions and network dynamics, contributing to a deeper understanding of how social networks evolve and influence social phenomena (Ripley et al., 2011).

5.14.2 Rate Function

The rate function is used to describe the probability that an actor will make a particular change to their network ties. Let *t* represent time, and let *S(t)* be the state of the network at time *t*, where *S(t)* is a matrix that represents the network ties between actors. Let *i* represents the actor who is making the change, and let *j* represents the actor whose tie is being changed. Let *ΔS* be the change that the actor *i* is considering making to their tie to *j*. The rate function is given by:

$$\lambda\left(i, j, \Delta S, \middle| S(t)\right) \tag{5.65}$$

where the actor *i* will choose to make the change *ΔS* to their tie to actor *j*, given the current state of the network at time *t*.

The rate function can depend on various factors, such as the attributes of the actors and the current state of the network. For example, the rate function might be higher for actors with certain attributes or for certain types of network changes. Let's say we have a social network of 5 individuals, and the rate function specifies that each actor's rate of changing their relationships with others depends on their degree centrality, which is a measure of how many connections an actor has. If actor A has a high degree centrality, they are more likely to make a move, while actor B with a low degree centrality is less likely to make a move. The rate function for actor A might be $\lambda(A, j, t) = 0.1$, while for actor B, it might be $\lambda(B, j, t) = 0.05$.

Once the rate function is specified, it can be used to simulate the evolution of the network over time. At each time step, each actor considers making a change to one of their network ties, and the probability of each possible change is given by the rate function. The model then randomly chooses one of the changes to make, based on the probabilities given by the rate function. This process is repeated for many time steps, and the resulting network is a simulation of the evolution of the real network over time. As a result, the rate function determines the probability of an actor changing one of their relationships within the network at a specific point in time. It specifies the frequency with which individual actors can make a move.

5.14.3 Objective Function

The objective function measures the attractiveness of different network configurations to each actor, taking into account their individual preferences and the utilities associated with the set of moves they can make. The objective function is the key component that governs how an actor decides to change its relationships within the network. This function consists of three elements: an evaluation function, an endowment function, and a random component.

The evaluation function reflects the actor's satisfaction with different possible network configurations. It can take into account various attributes of the actor, such as its position within the network, the characteristics of its alters, and the overall structure of the network. The evaluation function guides the actor to make changes that improve its perceived satisfaction with the network.

The endowment function represents the benefits that the actor gains from making particular changes. These benefits can be related to the actor's attributes, such as its social capital or its ability to access resources. The endowment function provides an incentive for the actor to make changes that will increase its benefits.

Finally, the random component captures the effects of unmeasured factors that influence the actor's decision-making process. These factors can include random events, such as changes in the external environment, or unobserved attributes of the actor or its alters. The random component adds a level of uncertainty to the model, which reflects the unpredictability of social interactions in the real world.

Together, these three components of the objective function provide a comprehensive framework for understanding how actors make decisions about changing their relationships within a network. By incorporating both observable and unobservable factors, the model captures the complexity of social dynamics and helps us to better understand how networks evolve over time.

The objective function can be written using mathematical notation as follows:

$$f_i\left(\beta,x\left(i \rightarrow j\right)\right)+g_i\left(\gamma,x,j\right)+\varepsilon\left(t,x,j\right) \tag{5.66}$$

The first part, $(\beta, x(i \rightarrow j))$, is the evaluation function, which reflects an actor's satisfaction with different network configurations. It is a function of the actor's attributes, represented by $x(i\sim>j)$, and the change in the network, represented by $i\sim>j$. The parameter β determines the strength of the relationship between the actor's attributes and their satisfaction with the network configuration.

The second part of the objective function is $g_i(\gamma, x, j)$, which is the endowment function. It represents the benefits gained from making particular changes to the network. It is a function of the actor's attributes, represented by x, and the network configuration, represented by j. The parameter γ determines the strength of the relationship between the actor's attributes and the benefits of making a particular change.

The third part of the objective function is $\varepsilon(t, x, j)$, which is the random component. It captures the effects of unmeasured factors that may influence an actor's decision to make a change in the network. It is a function of time t, actor attributes x, and network configuration j.

5.14.4 Method of Moments

The method of moments (MoM) estimator is a popular approach for estimating the parameters (θ) of a Stochastic Actor-Oriented Model (SAOM). It allows us to make inferences about the underlying mechanisms driving the evolution of a network (Zinilli, 2016). In this context, the estimation is performed based on the initial observation of the network ($x(t_0)$), which is treated as a given starting point. The advantage is that we don't need to make any assumptions about how $x(t_0)$ was generated. The MoM estimator relies on the idea that the expected values of certain statistics are related to the model's parameters. By equating the expected values of these statistics to their observed sample counterparts, we can estimate the parameters θ. The selection of appropriate statistics is crucial. They should be sensitive to changes in the parameters, meaning that higher values of a parameter should lead to higher expected values of the corresponding statistic.

The MoM estimator, therefore, involves solving a system of moment equations, where the expected values of the chosen statistics should match their observed values. By iteratively updating the parameter estimates based on simulated network evolution and the Robbins–Monro step, we can obtain the final estimates of the SAOM parameters. The Robbins–Monro step is a key component of the stochastic approximation algorithm used in estimating the parameters of the SAOM through the method of moments. It is named after Herbert Robbins and Sutton Monro, who introduced this iterative method in 1951 (Robbins & Monro, 1951). The Robbins–Monro step is used to update the parameter estimates in each iteration of the algorithm, bringing them closer to the true parameter values. In their influential paper, Robbins and Monro (1951) addressed the problem of estimating the unknown zero, denoted as θ_*, of a function $h: R^p \rightarrow R$. Although $h(\theta)$ is unknown, it can be estimated without bias using a function H that depends on a random variable ξ, satisfying $E\xi(H(\theta, \xi)) = h(\theta)$ for a fixed θ in the parameter space $\Theta \subseteq R^p$. The Robbins–Monro algorithm iteratively estimates the value of θ_* using observations $\xi_1, \xi_2, \ldots$, according to the following update rule:

$$\theta_n = \theta_{n-1} - \gamma_n \left[\mathrm{H}(\theta_{n-1}) + \xi_n \right] \tag{5.67}$$

Here is a breakdown of the components in the update rule.

θ_n and θ_{n-1}. The parameter estimates at the current iteration n and the previous iteration $n-1$, respectively. γ_n indicates the step size or learning rate at iteration n. It is chosen such that the sum of squared step sizes, $\sum \gamma_i^2$, is finite, while the sum of step sizes, $\sum \gamma_i$, is infinite. In other words, γ_n decreases over iterations to ensure convergence. Finally, $[\mathrm{H}(\theta_{n-1}) + \xi_n]$ is the value of the function $h(\theta)$ using the parameter estimate θ_{n-1} and the observed value ξ_n. The algorithm starts with an initial estimate θ_0 and iteratively updates the estimate using the observed data and the function H. The step size γ_n controls the size of the update and gradually reduces to allow for more precise estimation as the iterations progress.

The Robbins–Monro algorithm is designed to converge to the true value of θ, provided certain conditions are met, such as the continuity of $h(\theta)$ and the existence of a unique zero θ in the parameter space Θ. The convergence is typically assessed by analyzing the behavior of the estimated parameters over iterations. The use of a diminishing step size γ_n and the unbiased estimation of $h(\theta)$ through H allows for efficient estimation of the unknown parameter θ_*.

The Robbins–Monro algorithm, incorporating the step updates, simulation of network evolution, and convergence checks, allows for efficient estimation of the SAOM parameters even when explicit computation of the likelihood function is infeasible.

Let $\theta = (\rho_1, \ldots, \rho_m, \beta_1, \ldots, \beta_k)$ be a parameter vector with dimension $P = (M + K)$ for a specific SAOM specification. Estimating θ helps us understand the driving mechanisms behind network evolution. The estimation is based on the first observed network state, denoted as $x(t_0)$, which is considered given and not explicitly modeled. Therefore, we do not need to make any assumptions about the mechanisms that generated $x(t_0)$. Moreover, the estimated parameters only pertain to the network's evolution within the time interval $[t_0, t_M]$. Various estimation procedures can be used to estimate θ. Snijders (2001, 2010) proposed the method of moments (MoM) and maximum likelihood estimation (MLE) procedures. However, due to the SAOM's likelihood function being computationally infeasible to compute explicitly, MLE estimations are time-consuming. As a result, the MoM is the commonly used method in practice (Amati et al., 2015, 2019; Snijders, 2001).

The MoM is based on the concept that the expected values of statistics depend on the distribution's parameters. Suppose X is a random variable with a probability distribution determined by parameter $\theta \in R^P$. Our goal is to estimate θ. The MoM requires a P-dimensional vector of statistics $S = (S_1, \ldots, S_p)$, where each statistic S_p corresponds to a parameter θ_p $(1 < p < P)$.

The estimator $\hat{\theta}_{MoM}$ is calculated by equating predicted values to sample values:

$$E_\theta\left[S\right] = s \rightarrow s = \left(s_1, s_2, \cdots, s_P\right) \quad (5.68)$$

The selection of statistics S_P typically involves a formal method like sufficiency reduction. However, in the case of estimating SAOM parameters, this method is not applicable due to the complexity of the model. Hence, statistics are chosen heuristically, ensuring their sensitivity to the parameter θ. Heuristically means that the selection is based on practical considerations, intuition, or informal rules of thumb, rather than a formal and rigorous mathematical procedure. This strategy provides a practical and pragmatic way to make progress or approximate solutions in the absence of a formal method. Choosing statistics heuristically with the aim of ensuring their sensitivity to the parameter θ involves selecting statistics that are likely to reflect changes in θ, enabling better estimation and understanding of the impact of parameter on the system or model being studied. This sensitivity implies that higher values of θ_P lead to higher expected values of the statistic S_P, as expressed by the condition:

$$\frac{\partial}{\partial \theta_p} E_\theta\left[S_p\right] > 0 \quad (5.69)$$

Specifically, Snijders (2001) proposed using the statistic S_{ρ_m}, which captures the frequency at which actors make changes. For this parameter, a relevant statistic is the Hamming distance between the current observation $X(t_m)$ and the previous observation $X(t_{m-1})$:

$$S\rho_m = \sum_{i=1}^{n} \sum_{j=1}^{n} \left| X_{ij}\left(t_m\right) - X_{ij}\left(t_{m-1}\right) \right) \quad (5.70)$$

Furthermore, if a parameter β_k is assumed to be constant for all M periods, an adaptation of the prescription is used. Functions $S_k(x)$ are employed, with observed values $S_k(x(t_m))$, for $m = 0, 1, \ldots, M$. The MoM equation for these S_k is

$$\sum_{m=1}^{M} E_\theta\left[S_k\left(X\left(t_m\right)\right) \middle| X\left(t_{m-1}\right) = x\left(t_{m-1}\right)\right] = \sum_{m=1}^{M} S_k\left(x\left(t_m\right)\right) \quad (5.71)$$

For a given k, higher values of β_k are expected to lead to higher expected values of the corresponding function $s_{ik}(X(t))$ for $t > t_0$ and all $i \in N$. Thus, a sensitive statistic for the parameter β_k is

$$S_k\left(X\left(t_m\right)\right)=\sum_{i=1}^{n} s_{ik}\left(X_{ij}\left(t_m\right)\right) \tag{5.72}$$

In summary, the estimator $\hat{\theta}_{\text{MoM}}$ for θ is the solution of the system of $M + K$ moment equations:

$$\begin{cases} E_\theta\left[S_{\rho_m} \middle| X\left(t_{m-1}\right)=x\left(t_{m-1}\right)\right] & m=1,2,\ldots,M \\ \sum_{m=1}^{M} E_\theta\left[S_k\left(X\left(t_m\right)\right) \middle| = X\left(t_{m-1}\right)=x\left(t_{m-1}\right)\right]=\sum_{m=1}^{M} s_k\left(x\left(t_m\right)\right) & k=1,2,\ldots,K \end{cases} \tag{5.73}$$

where s_{ρ_m} is the observed value of S_{ρ_m}.

To approximate the covariance matrix of $\hat{\theta}_{\text{MoM}}$, the delta method can be applied (Amati et al., 2015, 2019). The delta method is used to approximate the variance of a function of random variables. It provides a way to estimate the variance of a parameter when the exact distribution of that quantity is unknown or difficult to compute. The covariance matrix is given by:

$$\sum\nolimits_{\hat{\theta}_{\text{MoM}}}=D_\theta^{-1}\sum\nolimits_S\left(D_\theta^{-1}\right) \tag{5.74}$$

where D_θ is the matrix of partial derivatives of the expected values of the statistics:

$$D_\theta=\left(\frac{\partial}{\partial\theta}E_\theta\left[S\right]\right) \tag{5.75}$$

where S represents the covariance matrix of the statistics. This formulation allows for the estimation of the parameter θ in the SAOM using the MoM, and the covariance matrix can be approximated for further analysis. The approach described in the text is an iterative stochastic approximation algorithm used to estimate the parameter θ in the SAOM. It entails network evolution simulations and parameter value updates based on the Robbins–Monro phase described previously. The algorithm, when applied to the SAOM environment, covers three phases:

Phase 1: In this initial phase, a small number of simulations is performed to obtain a rough estimate of the matrix of partial derivatives (D_θ) in Eq. (5.75). The derivatives can be estimated using either the finite difference method or the score function method. The initial parameter value (θ_0) is updated using the Newton–Raphson step:

$$\theta_1=\theta_0-a_0D_0^{-1}\left(\overline{S}-s\right) \tag{5.76}$$

where S is the mean of the simulated values of the statistics.

Phase 2: This phase consists of several sub-phases (usually four) with a fixed number of iterations. Each sub-phase involves simulations of network evolution and updates of the parameter value according to the Robbins–Monro step. The number of iterations is bounded by minimum and maximum values. A sub-phase ends when either the upper bound is reached or the first-order autocorrelation of the sequence S_i (values of statistics) is negative for all i. At the end of each sub-phase, the value of θ is estimated using the mean of the sequence θ_i.

Phase 3: In this phase, network evolution is simulated a large number of times using the value of θ determined at the end of Phase 2. The simulations are used to estimate the covariance matrix of the estimator $\hat{\theta}_{\text{MoM}}$ and to check the convergence of the algorithm. The matrix of partial derivatives D_θ is estimated as in Phase 1, while the covariance matrix S is estimated using direct Monte Carlo approximation. Convergence is checked based on the deviations between simulated values of the statistics and their observed values. The convergence is considered excellent when the ratios between the mean simulated values and their standard deviations are all below a certain threshold (0.1 or 0.15) (Amati et al., 2015, 2019). In the proposed estimation procedure, more statistics are used than parameters, resulting in an over-identified system of equations. Therefore, the regular method of moments (MoM) cannot be directly applied. Another approach employs the generalized method of moments (GMoM), which is a technique commonly used in econometrics and potentially offers greater efficiency than MoM. Unlike MoM, which sets the differences between expected values and sample counterparts to zero, GMoM involves minimizing a quadratic function that captures these differences. This approach allows to solve the minimization problem and estimate the parameters more effectively.

A modified version of the Robbins–Monro method, a stochastic approximation approach used for parameter estimation, is employed to develop the GMoM. The modified algorithm iteratively updates the parameter value based on the GMoM criterion, aiming to find the parameter value that minimizes the quadratic function. By incorporating more statistics and employing the GMoM with the adjusted Robbins–Monro algorithm, this estimation procedure leverages additional information and following Amati et al. (2015, 2019), this new approach has the capacity to provide more efficient parameter estimates compared to the traditional MoM approach. The generalized method of moments (GMoM) is employed when estimating a model that includes more statistics (moment conditions) than parameters. As explained above, in the traditional method of moments (MoM), the estimation is based on setting the sample moments equal to their corresponding population moments, effectively solving a system of equations. However, in some cases, the system becomes over-identified, meaning there are more equations than unknowns, making it impossible to find a solution by setting all differences to zero. To address this issue, GMoM extends the MoM by formulating the estimation as a minimization problem. Instead of setting the differences between sample moments and population moments to zero, GMoM minimizes a quadratic function that measures the weighted distance between these differences. The weights reflect the precision of the moments and can be based on theoretical considerations or estimated from the data. In the context of SAOM, where the configuration of ties between actors evolves over time, GMoM can be particularly useful. By considering the information from the previous time step, additional statistics can be included in the estimation procedure to improve the accuracy of parameter estimation. These additional statistics capture the dynamics of network evolution and provide valuable insights into how certain configurations originate. To minimize this quadratic function, a Q-dimensional vector of statistics, denoted as $S^* = (S_1, S_2, \ldots, S_Q)$, is used. These statistics are chosen based on their relevance to the estimation of the parameters. The differences between the expected values of these statistics and their sample counterparts are squared and weighted by a QxQ matrix W, known as the concentration matrix. The choice of W is important to ensure that the quadratic form is a meaningful measure of the distance. The GMoM estimator, θ_{GMoM}, is the value that minimizes this quadratic function. It is obtained by solving the minimization problem for the chosen statistics. However, estimating θ_{GMoM} analytically or numerically can be challenging, especially in the context of the SAOM where the network evolution is observed at discrete time points. To overcome these challenges, a modified version of the Robbins–Monro algorithm is employed. The algorithm aims to find the root of the equation that characterizes the minimization problem. This involves computing the first-order derivatives of the statistics with respect to the parameters and constructing a matrix B that approximates the product of the transpose of these derivatives and the concentration matrix. By solving the next Eq. (5.77), an approximation of the GMoM estimator θ_{GMoM} can be obtained:

$$B\left(E_\theta\left[S^*\right]-s^*\right)=0 \tag{5.77}$$

The matrix B serves as the normal equations for the estimation process and assigns weights to each statistic based on their sensitivity to parameter perturbations and their variance. This allows for the differential importance of each statistic in the estimation process.

As mentioned before, in the context of the MoM algorithm, the stochastic approximation method is employed in three phases. The GMoM algorithm also employs the stochastic approximation method, but with some modifications compared to the MoM algorithm. In the GMoM algorithm, the matrix of weights, denoted as W, is estimated in addition to the matrix of partial derivatives. The estimation of W is based on the inverse of the covariance matrix of the simulated statistics. The GMoM parameter estimation then proceeds similarly to the MoM algorithm, with the stochastic approximation method used in the second phase to update the parameter estimate based on the observed data and the estimated matrices. Overall, the stochastic approximation method is a common element in both the MoM and GMoM algorithms. It enables iterative updates of the parameter estimate based on observed data and estimated matrices. The specific differences lie in the estimation of the matrices and the inclusion of the matrix of weights in the GMoM algorithm, which allows for a more generalized and flexible parameter estimation (to get more details, see Amati et al., 2015, 2019; Snijders, 2017; Ceoldo et al., 2023).

5.14.5 R Code and Output

```
# Loading libraries
> library(RSiena)
> library(sna)

# Generating random adjacency matrices
> set.seed(1234)
> net1 <- matrix(rbinom(80^2, 1, 0.2), nrow = 80)
> net2 <- matrix(rbinom(80^2, 1, 0.3), nrow = 80)
> net3 <- matrix(rbinom(80^2, 1, 0.4), nrow = 80)

# Generating random actor attributes
> alcohol <- matrix(sample(0:2, 80*3, replace = TRUE), nrow = 80, ncol = 3)
> gender <- factor(sample(c("Male", "Female"), 80, replace = TRUE))

# Computing Moran's Index for the generated networks
> moran1 <- nacf(net1, alcohol [, 1], lag.max = 1, neighborhood.type = "out", type = "moran", mode = "digraph")
> moran2 <- nacf(net2, alcohol [, 2], lag.max = 1, neighborhood.type = "out", type = "moran", mode = "digraph")
> moran3 <- nacf(net3, alcohol [, 3], lag.max = 1, neighborhood.type = "out", type = "moran", mode = "digraph")
> moranInd <- c(moran1[2], moran2[2], moran3[2])
> moranInd

# Formatting the data
> friendship <- sienaNet(array(c(net1, net2, net3), dim = c(80, 80, 3)))
> alcoholbeh <- sienaNet(alcohol , type = "behavior")
> sex <- coCovar(as.numeric(gender) - 1)  # Convert gender to numeric (0 for Male, 1 for Female)
> myCoevData <- sienaDataCreate(friendship, alcoholbeh, sex)

# Specifying the effects
> myCoevEff <- getEffects(myCoevData)

# Running the model
> mymodel <- sienaModelCreate(useStdInits = FALSE, projname = "My_project")
> coevModel <- siena07(mymodel, data = myCoevData, effects = myCoevEff, batch=FALSE, verbose=TRUE, useCluster =
TRUE, nbrNodes=3, returnDeps=T)

Output

      Estimates, standard errors and convergence t-ratios

                                                        Estimate   Standard     Convergence
                                                                     Error        t-ratio
      Network Dynamics
        1. rate constant friendship rate (period 1)    92.2813 (   11.3662 )     0.2205
        2. rate constant friendship rate (period 2)   696.0731 ( 1928.5790 )    -3.6712
        3. eval outdegree (density)                    -0.3175 (    0.0178 )    -0.2146
        4. eval reciprocity                             0.0052 (    0.0345 )    -0.7389

      Behavior Dynamics
        5. rate rate alcoholbeh (period 1)               23.7835 (   18.4550 )    -0.3049
        6. rate rate alcoholbeh (period 2)               34.2060 ( 1910.5458 )     0.3490
        7. eval alcoholbeh linear shape                  -0.0511 (    0.0681 )    -0.7008
        8. eval alcoholbeh quadratic shape                0.8059 (    0.2254 )    -0.1120

      Overall maximum convergence ratio:     3.9132

# Splitting the plot window into two sections with a smaller window size
> par(mfrow = c(1, 2), mar = c(4, 4, 2, 2))

# Plotting the parabolic curve in dark orange
> parab <- function(x) {
  return(x^2)
}
> curve(parab(x), -0.5, 2, frame.plot = FALSE, axes = FALSE, ylab = "", xlab = "", ylim = c(-1.5, 4), xlim = c(-0.5, 2), lwd = 2,
col = "#FF8C00")
> axis(1, pos = c(0, 0), -1:8, col = "black", label = TRUE)
> axis(2, pos = c(-0.5, 0), c(-2, 4.5), labels = c("", ""), col = "black")
> mtext(expression(z[i] - bar(z)), side = 1, adj = 1)
> mtext(expression(f[i]^{beh}), adj = 0)

# Plotting the barplot of alcohol behavior distribution in dark orange
> distr <- t(apply(alcohol, 2, table) / 80)
> colnames(distr) <- c("no", "occasional", "regular")
> barplot(distr, beside = TRUE, ylim = c(0, 1), col = "#FF8C00", density = c(10, 30, 50), cex.axis = 0.8)
```

Output

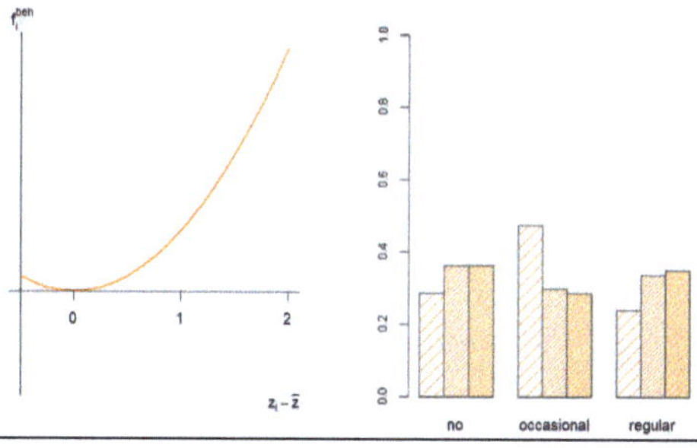

To evaluate the reliability of the simulation based on the convergence ratios, we can examine the individual *t-ratios* and the overall maximum convergence ratio as outlined in the criteria. The individual *t-ratios* represent the convergence of each estimated parameter at the individual level. According to the provided estimates, all individual *t-ratios* are larger in magnitude than 0. This indicates that the estimated parameters have not converged adequately at the individual level. The overall maximum convergence ratio reflects the convergence of the entire model. It should be less than 0.25 to indicate satisfactory convergence. The given overall maximum convergence ratio is 3.9132, which exceeds the acceptable limit of 0.25. This suggests convergence issues and raises concerns about the reliability of the simulation results. Obviously, these are generated data and not real data, the specific result of the estimates and convergence criteria are not important in this context. However, what is important is the procedure that was presented. The two pictures represent different aspects of the analysis. The parabolic curve represents the behavioral objective function. It is a visual representation of a parabolic eq. $2.8464{*}x^2{-}3.5464{*}x$, where x represents a variable or parameter. The curve helps visualize the relationship between the behavioral objective function and the variable x. The curve represents a convex relationship between the behavioral objective function and x. It is upward-opening, indicating that the rate of increase of the objective function slows down as x increases or decreases. The maximum point on the curve represents the optimal value of the behavioral objective function. This point indicates the value of x at which the objective function is maximized. The curve suggests that the objective function reaches its highest value at point 2.

The barplot represents the distribution of alcohol behavior across different waves or time points. Each bar in the plot represents the proportion or frequency of individuals in different categories of alcohol behavior (e.g., "no", "occasional" and "regular"). The barplot helps visualize the changes or distribution of alcohol behavior over time.

5.15 Case Study Using R: Co-Patent Network Using a Temporal Exponential Random Graph Model

In this application, we examine the influence of large metropolitan areas (LMAs) and explore the structural mechanisms of networks in driving innovation and economic growth within the European Union (EU). According to the definition provided by the OECD, LMAs are characterized as cities with a population exceeding 1.5 million people. The central question addressed is whether LMAs are becoming exclusive enclaves of innovators or if they are functioning as central nodes for the diffusion of knowledge and innovation throughout the EU. This analysis is crucial because it has significant implications for local economic development, the trajectory of economic convergence, and ultimately, the overall cohesion within the EU.

Furthermore, when investigating the formation of innovation networks, the concepts of preferential attachment and transitivity play a significant role in determining the network's final structural and functional characteristics (Glückler, 2007; Liu et al., 2015). Barabási and Albert's seminal study (1999) introduced preferential attachment as the fundamental mechanism responsible for the formation of scale-free networks. Essentially, preferential attachment suggests that the likelihood of forming new links increases with the number of existing links a node has. In the case of regional innovation networks, which undergo continuous growth through the addition of new nodes and links, observing preferential attachment processes allows us to comprehend the network's true collaborative innovation topology. In many real networks, such as those in knowledge and innovation systems, preferential attachment and growth coexist. In the context of regional co-invented patent networks, preferential attachment can be attributed to agglomerative factors. It is widely acknowledged that population and economic activity tend to concentrate spatially. Agglomeration refers to the concentration of economic activity and resources in specific regions or areas.

Transitivity, also known as triadic closure, denotes the likelihood that partners of one node will become partners with another, with triangles in a network used to observe this. Triangles are frequently seen as markers of social capital in knowledge and innovation networks (Coleman, 1988), establishing trust between nodes and enticing involvement in shared goals. Ter Wal (2014), for instance, showed the importance of transitivity in the development of a biotech network based on shared patents, suggesting that as a network grows and becomes more linked, the importance of geographic proximity may decrease, and the propensity for triadic closure may become a more effective mechanism for fostering long-distance collaborative links.

Within the context of a Temporal Exponential Random Graph Model (TERGM), "link memory" refers to the tendency of network ties to form or persist based on past tie formations within the network. This concept captures the idea that the formation of new ties is influenced by the existing network structure and the historical record of past tie formations. In a study on patent collaboration in a specific technological domain, Bauer et al. (2022) found that inventor pairs were more likely to continue working together if they had previously collaborated on patents, suggesting that prior productive partnerships build a foundation of trust, knowledge, and synergy between the inventors, which in turn encourages continuous cooperation in their future projects.

From a policy standpoint, innovation and knowledge transmission are crucial for fostering sustainable economic growth by enabling less inventive regions to catch up. However, it is significant to highlight that many public policies adopted by the EU to promote innovation and ease its spread have unintentionally contributed to a divergence between more developed and less advanced countries. Some studies, such as those by Amoroso et al. (2018), Barrios et al. (2019), and Filippetti and Zinilli (2023), have highlighted this unintended consequence.

To investigate the evolution of innovation networks within the EU, co-patenting data is used as a proxy for measuring innovation collaboration among 248 European regions over the period from 2012 to 2016. A Temporal Exponential Random Graph Model (TERGM) is employed to model the dynamics of innovation networks over time. Our analytic approach consists of two main steps: initially analyzing our networks by considering preferential attachment, transitivity, and link memory, and subsequently employing a TERGM to analyze these three structural statistics alongside other external factors, including the influence of LMAs. By applying this analytical approach, the study aims to shed light on the patterns of innovation collaboration and the flow of knowledge within the EU. This knowledge can inform policymakers and stakeholders about the effectiveness of current policies and initiatives related to innovation and regional development. Overall, the goal is to gain a deeper understanding of how LMAs function in the context of innovation and knowledge diffusion, and how this dynamic impacts economic development and cohesion within the EU.

Data and Variables

The dependent variable is the presence or absence of a co-patent linkage between two European regions. A linkage exists when at least two inventors from different regions are reported in the same patent document. The model includes both endogenous (network-based) and exogenous (characteristic-based) variables.

The endogenous variables are as follows:

- Edge: Controls for the number of edges in the network, analogous to the intercept term in a regression model.
- GWDegree: The generalized weighted degree is a measure used to quantify the degree or connectivity of a node in a network. GWDegree takes into account the weights or strengths associated with those links, and it can be considered a proxy of the preferential attachment mechanism.
- GWESP: Measures the tendency to form triangles or triads between three nodes, indicating transitivity or the shared partner structure.
- Edge memory: Reflects the stability of links over time, capturing the consistency or inconsistency of connections.

- Edge time: A time covariate testing for a time trend in the number of alliance edges over time, considering the financial crisis as a structural break.

  The exogenous variables are as follows:

- Country: A dichotomous variable indicating whether two regions in the same country collaborate significantly more than regions in different countries.
- LMA-region: A binary variable indicating whether a region is a Large Metropolitan Area (LMA). For this variable, we have used the nodemix function in an ERGM, this function models similarity between groups of nodes. We examine various combinations of LM-regions and regions, including:

  0-0 (Base outcome): this represents a link between two regions that are not classified as LMRs.
  0-1: this indicates a link between a region and an LMR.
  1-1: this denotes a link between two LMRs.

- GDP: regional gross domestic product per capita.
- GERD similarity: It stands for Gross Expenditure on Research and Development. It is a key indicator used to measure the total expenditure invested in research and development activities within each region. For this variable, we take the absolute difference in GERD between regions to capture the similarity. It quantifies the absolute disparity or gap in R&D expenditure between different regions.
- Geo proximity: The inverse of the distance in kilometers between the centroids of each region.
- EU: Dummy variable that represents whether a region is located within a country that is a member of the European Union (EU) or not.

Results

The results presented in this section are based on a bootstrapping sampling. Bootstrapping is a resampling technique that allows for the estimation of confidence intervals. These results provide insights into the estimated values of the TERGM variables and their corresponding bootstrapped means. The 95% confidence intervals indicate the range within which the true population values are likely to fall with a 95% level of confidence. As a result, if the 2.5% and 97.5% values of the confidence interval include zero, it indicates that the variable is not statistically significant. In other words, we cannot reject the hypothesis that the true value of the variable is zero. Therefore, we do not have sufficient evidence to claim that the variable has a significant effect.

Below is the code for constructing a TERGM on co-patenting networks between regions from 2012 to 2016.

Input

```
# Load required libraries
> library(igraph)
> library(network)
> library(btergm)

# Set working directory
> setwd("Path")

# Read data from a file
> data<-read.csv("Path/data.txt")

# Normalize the 'gerd' variable by dividing it by 'gdp'
> data$gerd <- data$gerd / data$gdp

# Split the dataset based on the 'year' variable
> split <- split(data, data$year)

# Extract data for each year
> years <- c("2012", "2013", "2014", "2015", "2016")

# Initialize empty lists outside the loop
> list_largemetro <- list()
> list_gdp <- list()
> list_gerd <- list()
> list_hhi_ipc <- list()
> list_EU <- list()
> list_dist <- list()

# Opening the loop
> for (year in years) {

# Create 'a' data frames
a_data <- split[[year]]
assign(paste("a", year, sep=""), a_data)

# Create 'pat' data frames
pat_data <- data.frame(a_data$source_L, a_data$target_L, a_data$X_freq)
assign(paste("pat", year, sep=""), pat_data)

# Create 'G' graph
G <- graph.data.frame(pat_data)
assign(paste("G", substr(year, 3, 4), sep=""), G)

# Convert 'G' graphs to adjacency matrices
graph <- get.adjacency(G, attr = 'a_data.X_freq')
assign(paste("graph", substr(year, 3, 4), sep=""), graph)

# Convert adjacency matrices to regular matrices and then to 'net' network objects
adjacency_matrix <- as.matrix(graph)
net <- network(adjacency_matrix, directed = FALSE)
assign(paste("net", substr(year, 3, 4), sep=""), net)
```

```
# Remove duplicated rows based on the 'source_L' column
duplicate <- a_data[!duplicated(a_data$source_L), ]
assign(paste("duplicate", substr(year, 3, 4), sep=""), duplicate)

# Extract attribute values
largemetropolitan <- duplicate$largemetropolitan
assign(paste("largemetropolitan", substr(year, 3, 4), sep=""), largemetropolitan)

gdp <- duplicate$gdp
assign(paste("gdp", substr(year, 3, 4), sep=""), gdp)

gerd <- duplicate$gerd
assign(paste("gerd", substr(year, 3, 4), sep=""), gerd)

hhi_ipc <- duplicate$hhi_ipc
assign(paste("hhi_ipc", substr(year, 3, 4), sep=""), hhi_ipc)

EU <- duplicate$EU
assign(paste("EU", substr(year, 3, 4), sep=""), EU)

# Set vertex attribute 'largemetropolitan' for each 'net' network
set.vertex.attribute(net, paste("largemetropolitan", substr(year, 3, 4), sep=""),
  largemetropolitan)

# Set vertex attribute 'gdp' for each 'net' network
 set.vertex.attribute(net, paste("gdp", substr(year, 3, 4), sep=""), gdp)

# Set vertex attribute 'gerd' for each 'net' network
set.vertex.attribute(net, paste("gerd", substr(year, 3, 4), sep=""), gerd)

# Set vertex attribute 'hhi_ipc' for each 'net' network
set.vertex.attribute(net, paste("hhi_ipc", substr(year, 3, 4), sep=""), hhi_ipc)

# Set vertex attribute 'EU' for each 'net' network
set.vertex.attribute(net, paste("EU", substr(year, 3, 4), sep=""), EU)

# Create 'dist' data frames
dist_data <- data.frame(a_data$source_L, a_data$target_L, a_data$distance_std)
assign(paste("dist_data", substr(year, 3, 4), sep=""), dist_data)

# Create 'Dis' graphs
Dis <- graph.data.frame(dist_data)
assign(paste("Dis", substr(year, 3, 4), sep=""), Dis)

# Convert 'Dis' graphs to adjacency matrices
Dis_g <- get.adjacency(Dis)
assign(paste("Dis_g", substr(year, 3, 4), sep=""), Dis_g)

# Convert adjacency matrices to regular matrices
mat_dist <- as.matrix(Dis_g, attrname='myEdgeWeight')
```

```
# Set diagonal to 0
diag(mat_dist) <- 0

# Convert matrices to 'matdis' network objects
matdis <- network(mat_dist, directed = FALSE, ignore.eval=FALSE, names.eval='lag')
assign(paste("matdis", substr(year, 3, 4), sep=""), matdis)

# Add the variables to the lists generated before the loop
list_largemetro[[year]] <- largemetropolitan
list_gdp[[year]] <- gdp
list_gerd[[year]] <- gerd
list_hhi_ipc[[year]] <- hhi_ipc
list_EU[[year]] <- EU
list_dist[[year]] <- matdis

}

# Create a list of matrices based on the network
> allNet<-list(net12, net13, net14, net15, net16)
> names(allNet) <- c("2012", "2013", "2014", "2015", "2016")

# Iterate through each network in allNet and set vertex and network attributes
> for (i in 1:length(allNet)) {
  allNet[[i]] <- set.vertex.attribute(allNet[[i]], "largemetropolitan12", largemetropolitan12)
  allNet[[i]] <- set.vertex.attribute(allNet[[i]], "hhi_ipc12", hhi_ipc12)
  allNet[[i]] <- set.network.attribute(allNet[[i]], "mat_dist", mat_dist)
  allNet[[i]] <- set.vertex.attribute(allNet[[i]], "list_gdp", list_gdp)
  allNet[[i]] <- set.vertex.attribute(allNet[[i]], "list_gerd", list_gerd)
  allNet[[i]] <- set.vertex.attribute(allNet[[i]], "list_EU", list_EU)
}

# Fit the first btergm model (model1) to allNet and specified endogenous terms
> model1<- btergm(allNet~ edges+gwdegree(0.5, fixed = T)+gwesp(0.5,fixed=T)+memory(type
= "stability", lag=1)+timecov(transform = function(t) t),R=50)

# Print the summary of model1
> summary(model1)
```

Output

```
Estimates and 95% confidence intervals:
                    Estimate  Boot mean      2.5%     97.5%
edges              -2.402370  -2.374652  -2.9292   -1.9195
GWDegree          -12.835603 -12.895587 -14.0757  -11.6062
GWESP               0.949699   0.954861   0.8560    1.0411
Edge Memory         0.454794   0.453571   0.2740    0.5704
Edge Time          -0.053933  -0.059335  -0.1253    0.0611
```

The effect of Edges is −2.40, and it is statistically significant, as indicated by the confidence interval (−2.93 to −1.92) that does not include zero. The negative coefficient of Edges is a common observation in real networks, indicating that these networks tend to have lower density compared to exponential random networks. This suggests that the formation of network ties is not purely random but is influenced by various factors. Moving on to GWDegree, the estimated effect of −12.83 suggests a penalization of highly connected nodes, indicating that nodes with higher global weighted degree are less likely to form additional ties. This negative and significant coefficient implies the opposite of a preferential attachment mechanism—high-degree nodes are less likely to attract new ties, suggesting a possible saturation or competition effect. The positive and significant coefficient of GWESP reveals the importance of transitivity mechanisms in shaping the network. The presence of transitivity indicates that nodes tend to form connections with each other if they have common neighbors, leading to the formation of triadic relationships. This finding highlights the role of triadic closure as a driving force in the evolution of the regional co-invention networks.

Moving on to Edge Memory, the estimated effect of 0.45 indicates that there is a tendency of edges to persist over time. This suggests that once connections are formed, they are more likely to be maintained over time. Finally, we focus on the edge time parameter. The range (−0.12 to 0.06) provided by the confidence interval indicates the likely range for the real effect size of this variable. This indicates that the timing of tie formation or dissolution might influence network dynamics in a context-dependent way.

Input

```
# Fit the first btergm model (model2) to allNet and all terms
> model2<- btergm(allNet~ edges + gwdegree(0.5, fixed = T) + gwesp(0.5,fixed=T) + memory(type = "stability", lag=1) + timecov("largemetropolitan12", transform = function(t)t) + nodemix("largemetropolitan12") + nodecov("list_gdp") + absdiff("list_gerd") +edgecov(mat_dist) + nodecov("list_EU"),R=50)

# Print the summary of model2
> summary(model2)

# Calculate goodness-of-fit statistics for model2
> gof.model <- btergm::gof(model2, nsim = 100, target = allNet[[4]], formula = allNet~edges + gwdegree(0.5, fixed = T) + gwesp(0.5,fixed=T) + memory(type = "stability", lag=1) + timecov("largemetropolitan12", transform = function(t) t) + nodemix("largemetropolitan12") + nodecov("list_gdp") + absdiff("list_gerd") + edgecov(mat_dist) + nodecov("list_EU"), coef = coef(model2), statistics = c(esp, dsp, deg, rocpr))

# Plot the goodness-of-fit results with ROC and PR curves
> plot(gof.model, roc.rgraph = TRUE, pr.rgraph = TRUE)
```

Output

```
Estimates and 95% confidence intervals:
                               Estimate   Boot mean      2.5%     97.5%
Edges                       -1.1191e+01 -1.1283e+01 -12.6202   -9.5592
GWDegree                    -1.2275e+01 -1.2127e+01 -13.5584   -9.5093
GWESP                        8.9171e-01  8.8830e-01   0.8124    0.9740
Edge Memory                  4.1228e-01  4.0642e-01   0.2435    0.6000
Edge Time(LMR)              -7.0824e-02 -6.7249e-02  -0.1610    0.0261
mix.largemetropolitan.0.1    2.3522e-01  2.3165e-01   0.1240    0.3859
mix.largemetropolitan.1.1    4.8353e-01  4.6090e-01   0.1101    0.9386
Gdp                          4.4171e-01  4.4366e-01   0.3755    0.4901
GERD similarity             -2.4788e-01 -2.4742e-01  -0.5310   -0.0363
Geo. Proximity               3.2107e-08  3.6156e-08   0.0000    0.0000
EU                           8.4979e-03  4.0937e-02  -0.3175    0.5506
```

In the second model, the estimated effect of Edges is −11.32, and it is statistically significant. Moreover, also in this case, the GWDegree variable also shows a negative coefficient. However, this does not support the presence of preferential attachment; rather, it suggests a penalization of highly connected nodes, indicating that nodes with higher global weighted degree are less likely to form additional ties. As in the previous model, the GWESP variable is significance. The positive coefficient of GWESP emphasizes the importance of transitivity mechanisms in shaping the network. This finding highlights the role of triadic closure as a driving force in the evolution of the network, facilitating the clustering of inventors and the emergence of local communities within the network. Furthermore, we find that Edge Memory is significant, with a positive parameter. This implies a high degree of stability in the relationships between inventors over time. On the other hand, the interaction variable Edge Time*LMR is not found to be significant. In this second model, we also consider several exogenous variables that may contribute to the structure and dynamics of co-patent network. Moving on to the mix.largemetropolitan.0.1 variable, the estimated effect is 0.23, and the confidence interval of 0.12 to 0.39 includes zero, indicating non-significance. As far as our variables of interest are concerned, the variable Large Metropolitan Region (LMR) is positive, indicating a greater propensity of LMRs to connect within the overall network. The variable region, which is composed of three levels (0-0, 0-1 and 1-1), is always positive for the levels 0-1 (links between regions and LMRs) and 1-1 (links between LMRs), compared to the base outcome 0-0 (links between non LM-regions). LMR 0-1 (links between regions and LMRs) is positive and suggests that there is a greater propensity for connections between regions and large metropolitan regions. The positive coefficient of LMR 1-1 (links between LMRs) indicates that there is a higher likelihood of connections between different large metropolitan regions. This suggests that LMRs tend to establish collaborations or interactions with other LMRs within the network. Considering the GDP variable, the estimated effect is 0.44, with a confidence interval of 0.37 to 0.49 indicating statistical significance. This positive relationship suggests that regions with higher Gross Domestic Product (GDP) values have a stronger presence in the co-invention network. Higher GDP values often indicate a greater capacity for innovation, research and development activities, and access to resources. Such regions are more likely to attract skilled inventors and foster collaborations within the network. In contrast, the GERD similarity variable exhibits a negative estimated effect of −0.25, with a confidence interval of −0.53 to −0.04, indicating statistical significance. This variable represents the similarity in Gross Expenditure on Research and Development (GERD) between connected regions. The presence of a negative coefficient implies that regions with different research and development expenditures exhibit a reduced likelihood of establishing connections. This observation indicates a preference for linking regions possessing complementary research profiles, or a tendency to encourage collaborations that enhance the diversity of research endeavors within the network. Moving on to the geographical proximity variable, the estimated effect is very close to zero. The confidence interval suggests statistical significance, although the effect size is negligible. This indicates that the geographical proximity between connected regions is important but does not have a substantial impact on the formation of co-invention connections in this network. Other factors, such as technological advancements, communication networks, or shared research interests, may override the influence of geographical distance in shaping collaborations between inventors. The EU variable is the last to be considered, with an estimated effect of 0.008 and a confidence range of −0.32 to 0.55, which includes zero, suggesting a non-statistical significance. In the next box, you can find the code to assess the goodness-of-fit of the model 2 with both endogenous and exogenous variables. Goodness-of-fit evaluation helps determine how well the model captures the observed network structure.

Input

```
> gof.model <- btergm::gof(
model2,                          # Model object used for evaluation
nsim = 100,                      # Number of simulations for inference
target = allNet[[4]],            # Target network for comparison
formula = allNet ~ edges +       # Model formula specifying network terms.
gwdegree(0.5, fixed = T) +
gwesp(0.5, fixed = T) +
memory(type = "stability", lag = 1) +
timecov("largemetropolitan12", transform = function(t) t) +
nodemix("largemetropolitan12") +
nodecov("list_gdp") +
absdiff("list_gerd") +
edgecov(mat_dist) +
nodecov("list_EU"),
coef = coef(model2),    # Coefficients extracted from "model2" for simulation purposes
statistics = c(esp, dsp, deg, rocpr)    # Essential statistics used for evaluation and analysis
)

# Plot the results
> plot(gof.model, roc.rgraph = TRUE, pr.rgraph = TRUE)
```

Output

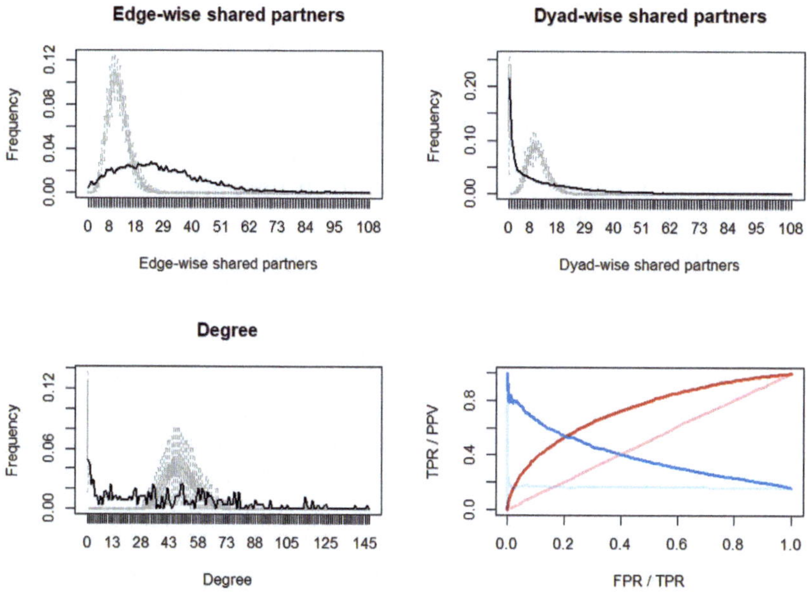

The goodness-of-fit (GoF) indicates that the model's fit to the observed network is not satisfactory; it suggests that the model might not accurately capture the underlying processes shaping the network structure. In such a case, it is important to interpret the results cautiously and consider limitations of the model. The discrepancies between the model's predictions and the observed network structure could highlight areas where the model fails to capture important dynamics or dependencies. It indicates that there are factors influencing the network formation that are not adequately accounted for by the model's current specifications. To further improve the model and enhance its goodness-of-fit, it may be necessary to revise the model's formulation, include additional relevant variables or covariates, or adjust parameter values.

5.15.1 Conclusions

The analysis reveals that the European innovation collaboration network has become increasingly integrated over time, with large metropolitan regions (LMRs) playing an essential role. These regions have not only strengthened their connections among themselves but have also expanded collaboration with regions outside their immediate network. This development challenges the notion that large metropolitan regions are becoming more exclusive and detached from others. In Europe, LMRs act as key hubs for collaborative innovation, simultaneously integrating peripheral regions and leading to a more distributed network of innovation. This phenomenon can be attributed to factors such as policy approaches, the polycentric urban structure of Europe, and the process of integration within the EU.

The integration process within the EU, particularly the enlargement involving Central and Eastern European countries, has reshaped the geography of production. The findings suggest a complex relationship between LMRs and other regions in terms of economic convergence. While the increasing centrality of LMRs may contribute to economic divergence, the integration of peripheral regions holds the promise for a convergence process.

The findings are promising with respect to the endogenous mechanisms shaping the co-patenting network among European regions. The results indicate that transitivity plays a significant role in the formation of ties, highlighting the importance of triadic closure—regions are more likely to collaborate if they share common partners. This tendency to cluster indicates that regions often build long-term partnerships in specific technological areas, forming regional communities of innovation. In contrast, the negative and significant coefficient for degree-related statistics suggests a departure from a pure preferential attachment dynamic. Rather than reinforcing existing hubs, the structure appears to discourage excessive concentration of ties around already well-connected regions. This may reflect capacity constraints or a strategic diversification of collaborations. Together, these mechanisms illustrate that European co-patenting networks are shaped both by local clustering dynamics and a tendency to balance centralization with broader connectivity.

The European innovation collaboration network's increased integration, with LMRs at the forefront, underscores their crucial role in driving collaborative innovation and fostering stronger connections both within and beyond their immediate regions.

5.16 Case Study Using R: A Friendship Network Using a Stochastic Actor-Oriented Model

This application uses the Stochastic Actor-Oriented Model (SAOM) to examine the friendship network among adolescents. The primary objective of this research is to investigate the organizational structure of the network and explore how network dynamics are influenced by individual characteristics. To achieve these goals, the R package "RSiena" is employed to analyze the effects of various features on the evolution of the network.

By employing the "RSiena" package, the research investigates how specific features of adolescents, such as gender, academic performance, and popularity, influence the formation of friendships within the network. The provided code snippet sets up the necessary data structures, loads the required packages, and defines the effects to be included in the analysis.

Adolescent friendship networks have been the subject of extensive research, with studies employing social network analysis to delve into their complex dynamics and the factors that shape them. In one study (Steglich et al., 2006), the focus was on actor-driven models that explored the coevolution of networks and individual characteristics. Researchers employed the SIENA software to assess the strength of homophily and assimilation processes. Interestingly, the findings revealed that connections in the network were generally avoided unless they offered desirable outcomes, such as reciprocation or transitive embeddedness.

Another study (Hawe & Ghali, 2008) investigated the relationships among teachers at a high school in Canada, seeking to uncover the relationship profiles within the network. The researchers discovered that the network structure primarily revolved around a select few dominant individuals, including the secretary, principal, and vice-principal, who played significant roles in shaping the overall dynamics of the network.

Long et al. (2019, 2020) focused on peer marginalization, peer withdrawal, and homophily associated with adolescent health status using social network analysis. In one study, they explored the relationship between anxiety disorders, disruptive behavior disorders, and friendships among adolescents. Surprisingly, while disruptive behavior disorders exhibited social aggregation and reward, anxiety disorders did not display similar tendencies. In a separate investigation, they examined the connections between adolescents with chronic health conditions and their friendships, finding no significant differences in the social network structure compared to peers without chronic illnesses.

In conclusion, these studies employ social network analysis to delve into the complexities of adolescent friendship networks and shed light on various influential factors. The findings highlight the significance of factors such as homophily, individual behaviors, parental influence, and the broader school environment in shaping the dynamics and structure of these networks. By deepening our understanding of these intricate social systems, researchers can pave the way for interventions and policies aimed at fostering positive peer relationships and promoting healthier lifestyles among adolescents.

The specific objective of this application is to investigate how the network is organized and to understand how the features of individuals within the network affect its evolution. The analysis includes steps such as creating dependent and covariate matrices, specifying effects to be evaluated, creating a data object, setting up the Siena algorithm, and performing the Siena analysis. Additionally, the code generates plots and tables to summarize the results and evaluates the goodness-of-fit of the model.

5.16.1 Data and Variables

The data used in this application consists of observations from 26 students across five waves. The exogenous variables included in the analysis are as follows:

Math: This variable represents the math scores of the students and is categorized into three levels: 1 for low math scores, 2 for medium math scores, and 3 for high math scores. It provides information about the academic performance of the students in the domain of mathematics.

Gender: This variable indicates the gender of the students. It is represented by a binary value, with 0 representing female students and 1 representing male students. Gender is a fundamental demographic characteristic that may influence social interactions and network dynamics.

Popularity: This variable measures the popularity of students based on the number of friends they have on Facebook. It is recorded for each of the five waves, providing insights into the students' social connections and their social status within the friendship network.

These variables—math, gender, and popularity—serve as the primary characteristics or attributes of the students within the friendship network. They are used to examine how these individual features influence the organization and evolution of the network over time. By considering these variables, your study aims to explore the interplay between academic performance, gender, popularity, and the dynamics of friendship networks among adolescents. The model includes similarity effects of exogenous variables. Specifically, it examines the similarity in three variables: *math scores*, *popularity*, and *gender*. These effects analyze how the similarity or dissimilarity in these variables between two students influences the likelihood of them forming or dissolving a friendship tie.

The endogenous variables are as follows:

Transitive triplets: Transitive triplets refer to patterns in the network where if actor A is connected to actor B, and actor B is connected to actor C, then there is a tendency for actor A to be connected to actor C as well. This effect captures the transitivity or clustering of relationships within the network.

Reciprocity: It examines the tendency for mutual friendships to form within the network. It investigates whether there is a higher likelihood of reciprocated friendship ties, where two students are friends with each other, compared to non-reciprocated ties where only one student considers the other as a friend.

Betweenness: Betweenness is a measure of the extent to which an actor lies on the shortest paths between other actors in the network. It quantifies the importance or centrality of an actor in facilitating communication or information flow between other actors. Including the betweenness effect allows for the examination of the influence of actor centrality on network dynamics.

By including these effects, we take into account the influence of transitive triplets and betweenness in the evolution of the friendship network. This helps to capture additional aspects of network dynamics and understand the mechanisms that shape the structure and relationships within the network.

5.16.2 Results

The next figure displays the friendship network of pupils across the five time points. The vertices in the network represent individual pupils, and their colors reflect their math scores. Specifically, vertices are colored red to represent low math scores, black to represent medium math scores, and green to represent high math scores. Pupils with missing math scores are represented by white vertices. By visualizing the network in this way, we can observe the distribution of math scores within the friendship network and hypothetically identify any patterns or clusters based on math performance.

Input

```
# Setting the working directory and loading the libraries
> setwd("path")
> library("RSiena")
> library("xtable")

# The following code provides data from several text files and assigns them to different variables:
> net1 <- as.matrix(read.table("net1.dat",na.strings="NA"))
> net2 <- as.matrix(read.table("net2.dat",na.strings="NA"))
> net3 <- as.matrix(read.table("net3.dat",na.strings="NA"))
> net4 <- as.matrix(read.table("net4.dat",na.strings="NA"))
> net5 <- as.matrix(read.table("net5.dat",na.strings="NA"))
> math<- as.matrix(read.table("math.txt",na.strings="NA"))
> gender <- as.matrix(read.table("gender.txt", na.strings="NA"))
> popularity <- as.matrix(read.table("Facebook.txt", na.strings="NA"))
> numberActors <- dim(popularity)[1]
> numberActors

# To inform RSiena that the adjacency matrices are network data and specify their order
> School<sienaDependent(array(c(net1,net2,net3,net4,net5), dim = c(numberActors,
numberActors, 5)))
> table(school,useNA='always')
> table(net1,net2,useNA='always')

# Light to dark for gender
> color3 <- c("#FF0000", "#000000", "green") g.layout <- gplot(apply(school, c(1,2), max),
usearrows=F)

# Set the layout of the plots to have one row and five columns
> par(mfrow=c(1,5))

# Adjust width and height as needed
> options(repr.plot.width = 1, repr.plot.height = 70)

# Function to create five plots
> gplot(net1, vertex.col=color3[math[,1]],vertex.cex = 2, edge.lwd = 1, arrowhead.cex=.75,
edge.col='gray22', coord=g.layout, main='Math score t1')
> gplot(net2, vertex.col=color3[math[,2]],vertex.cex = 2, edge.lwd = 1, arrowhead.cex=.75,
edge.col='gray22', coord=g.layout, main='Math score t2')
> gplot(net3, vertex.col=color3[math[,3]],vertex.cex = 2, edge.lwd = 1, arrowhead.cex=.75,
edge.col='gray22', coord=g.layout, main='Math score t3')
> gplot(net4, vertex.col=color3[math[,4]],vertex.cex = 2, edge.lwd = 1, arrowhead.cex=.75,
edge.col='gray22', coord=g.layout, main='Math score t4')
> gplot(net5, vertex.col=color3[math[,5]],vertex.cex = 2, edge.lwd = 1, arrowhead.cex=.75,
edge.col='gray22', coord=g.layout, main='Math score t5')
```

Output

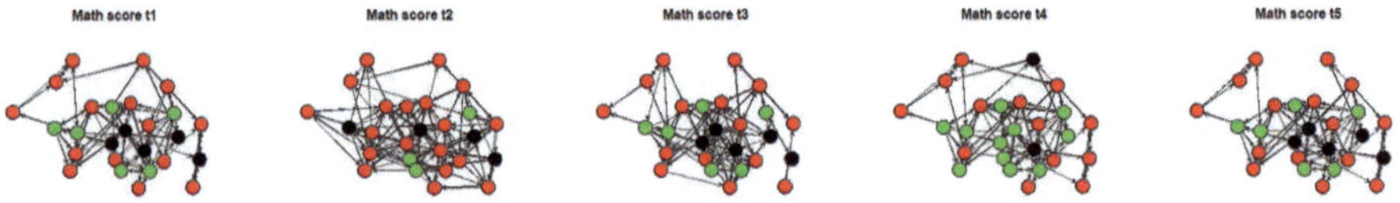

As observed at the first time point, math scores and popularity are used as changing covariates and gender is regarded as a constant covariate.

Input

```
# Define the covariates for the RSiena model
> math1 <- varCovar(math)             # varCovar is used to specify the changing covariates
> popularity1 <- varCovar(popularity)
> gender1 <- coDyadCovar(gender)      # coDyadCovar is used for the constant covariate

# Adjacency matrices and exogenous variables are combined to create a Siena data object
> Data <- sienaDataCreate(school,math1,popularity1,gender1)
> Data

# Specify the effects
> Effects <- getEffects(Data)
> effectsDocumentation(Effects)
> Effects <-includeEffects(Effects,TriadCensus, between)
> Effects <-includeEffects(Effects, include = TRUE, RateX , type = "rate", interaction1=
"math1")
> Effects <-includeEffects(Effects, include = TRUE, RateX , type = "rate", interaction1=
"popularity1")
> Effects <-includeEffects(Effects, include = TRUE, RateX , type = "eval", interaction1=
"gender1")
> Effects <-includeEffects(Effects, simX, interaction1="math1", type="eval")
> Effects <-includeEffects(Effects, sameX, interaction1="popularity1", type="eval")
> Effects <-includeEffects(Effects, simX, interaction1="gender1", type="eval")
> Effects

# We specify the type of model we are fitting
> Model <- sienaAlgorithmCreate(projname='school',firstg = 0.01, seed=745711)

# Estimating the model
> Model <- sienaAlgorithmCreate(projname='school',firstg = 0.01, seed=745711)
> Results <- siena07(Model,data=Data, effects=Effects, prevAns = school, batch=FALSE,
verbose=TRUE, useCluster=TRUE, nbrNodes=3, returnDeps=T)
> Results
```

Output

```
Estimates, standard errors and convergence t-ratios

                                      Estimate   Standard   Convergence
                                                   Error      t-ratio

Rate parameters:
  0.1       Rate parameter period 1   20.5318  ( 4.8941   )
  0.2       Rate parameter period 2    8.6307  ( 1.0599   )
  0.3       Rate parameter period 3    7.1255  ( 0.9699   )
  0.4       Rate parameter period 4    0.6990  ( 0.1823   )

Other parameters:
  1.  rate effect math1 on rate        0.0752  ( 0.1160   )    0.0655
  2.  rate effect popularity1 on rate  0.0000  ( 0.0003   )   -0.0530
  3.  eval outdegree (density)        -0.1976  ( 0.1655   )    0.0373
  4.  eval reciprocity                 1.2751  ( 0.1304   )    0.2149
  5.  eval betweenness                -0.3419  ( 0.0573   )    1.0009
  6.  eval gender1 similarity          0.3580  ( 0.0974   )    0.1099
  7.  eval math1 similarity            0.1101  ( 0.1121   )   -0.0940
  8.  eval same popularity1            0.5683  ( 0.8064   )    0.1015

Overall maximum convergence ratio:    2.0018


Total of 2159 iteration steps.
```

To evaluate the reliability of our simulation, we can examine the convergence ratios in the final column of the results. Specifically, we need to check two criteria: individual t-ratios and the overall maximum convergence ratio. Individual t-ratios should be less than the absolute value of 0.1. Looking at the provided estimates, all the individual t-ratios fall within acceptable limits as they are smaller in magnitude than 0.1 (except for reciprocity). This suggests that the estimated parameters have converged adequately at the individual level. The overall maximum convergence ratio should be less than 0.25. In the results, the overall maximum convergence ratio is reported as 2.0018, which significantly exceeds the recommended threshold. This indicates that there may be room for improvement in terms of convergence, and further refinement of the estimation process could be beneficial.

In summary, while the individual t-ratios meet the convergence criterion, the overall maximum convergence ratio exceeds the recommended threshold. This suggests that although the estimates at the individual level are reliable, there may be some lack of overall convergence in the simulation. It is advisable to investigate possible sources of convergence issues and consider refining the simulation or exploring alternative techniques to improve the overall convergence of the model.

Input

```
# To give approximate p-values
> parameter <- Results$effects$effectName
> estimate <- Results$theta
> st.error <- sqrt(diag(Results$covtheta))
> normal.variate <- estimate/st.error
> p.value.2sided <- 2*pnorm(abs(normal.variate),lower.tail=FALSE)
> data.frame(parameter, estimate=round(estimate,3), st.error=round(st.error,3),
normal.variate=round(normal.variate,2), p.value=round(p.value.2sided,4))
```

Output

```
                   parameter estimate st.error normal.variate p.value
1        effect math1 on rate    0.075    0.116           0.65  0.5171
2 effect popularity1 on rate    0.000    0.000          -0.07  0.9446
3        outdegree (density)   -0.198    0.165          -1.19  0.2323
4                reciprocity    1.275    0.130           9.78  0.0000
5                betweenness   -0.342    0.057          -5.96  0.0000
6          gender1 similarity    0.358    0.097           3.68  0.0002
7            math1 similarity    0.110    0.112           0.98  0.3262
8           same popularity1    0.568    0.806           0.70  0.4810
```

The analysis shows the effects of various variables on the rate parameter, which represents the rate at which friendship ties are formed or dissolved. The results revealed that math scores (math1) had no significant effect on the rate parameter, with an estimated effect of 0.075 and a non-significant p-value of 0.5171. This suggests that there is no strong evidence to support the idea that students with similar math scores are more likely to form or dissolve friendship ties. Similarly, the effect of popularity scores (popularity1) on the rate parameter was estimated to be close to zero, with a non-significant p-value of 0.9446. This indicates that popularity does not play a significant role in the dynamics of friendship networks. In contrast, certain network structural factors showed significant effects. Reciprocity, which measures the tendency for mutual friendships, had a positive effect on the rate parameter, with an estimated effect of 1.275 and a highly significant p-value. This suggests that the presence of reciprocal friendships increases the likelihood of forming or dissolving friendship ties. Additionally, betweenness centrality, which measures a student's influence in the network, had a significant negative effect on the rate parameter, with an estimated effect of −0.342 and a highly significant p-value. This implies that students with higher betweenness centrality are less likely to form or dissolve friendships. Furthermore, gender similarity (gender1 similarity) had a significant positive effect on the rate parameter, indicating that students of the same gender are more likely to form or dissolve friendships; the estimated effect was 0.358. Furthermore, the effects of math similarity (math1 similarity) and same popularity (same popularity1) were not statistically significant. Overall, these findings provide insights into the factors influencing the dynamics of friendship networks, highlighting the importance of reciprocal relationships, network centrality, and gender similarity.

Input

```
# Summary table of the results
> siena.table(Results, type='html', tstat=TRUE, d=2, sig=TRUE)
> save.image("results.txt")

# Time heterogeneity
> timetest <- sienaTimeTest(Results)
> summary(timetest)
> plot(timetest, effects=c(1,2,3,4,5,6,7))
```

Output

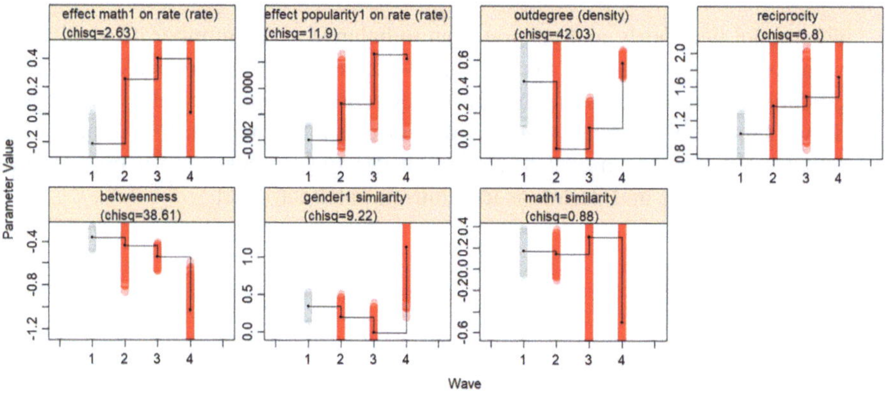

Input

```
# To perform goodness-of-fit testing in RSiena
> gof.id <- sienaGOF(Results, verbose=TRUE, varName="School", join=T, cumulative=F)
> plot(gof.id)
```

Output

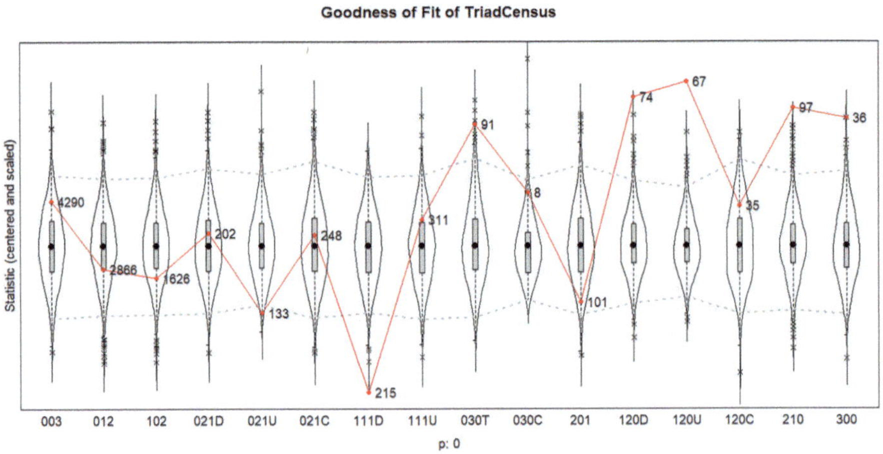

The model fits poorly and generates an imbalance in the number of triad types. There are several steps we can take to improve the model fit. There is likely a model specification issue in our situation. To better represent the dynamism of the network, we should think about including new effects in our model. It is important to note that improving the model fit can be an iterative process, and it may require multiple adjustments and iterations to achieve a satisfactory fit.

5.17 Conclusions

In conclusion, our application provides valuable insights into the dynamics of friendship networks among students. We found that academic factors, such as math scores and popularity, do not exert a significant influence on the formation or dissolution of friendship ties. This suggests that academic performance and social status may not be strong determinants of friendship dynamics among students. On the other hand, our results revealed important social and structural factors that shape friendship networks. Reciprocity, indicating mutual friendships, was found to have a positive impact on the likelihood of forming friendship ties. This highlights the significance of reciprocal relationships and mutual connections in driving the dynamics of friendships. Furthermore, our study uncovered the role of betweenness centrality, a measure of influence in the network, in friendship dynamics. Students with higher betweenness centrality, who act as important bridges between other students, were less likely to form friendships. This suggests that students in influential positions tend to have more stable friendship ties or experience fewer changes in their social connections. Moreover, the gender similarity effect emerged as an important factor in shaping friendship networks. Students of the same gender exhibited a higher likelihood of forming or dissolving friendships, indicating a preference for same-gender friendships among students. However, our study did not find a significant association between math similarity and friendship dynamics. This suggests that shared academic performance in math does not strongly influence the likelihood of forming friendship ties among students.

It is important to acknowledge that the results presented in this study serve as an application of the SAOM (Stochastic Actor-Oriented Model) framework to understand friendship dynamics among students. The primary goal of this study was to demonstrate the functionality and capabilities of the SAOM model rather than providing definitive conclusions about the specific context under investigation. The findings and interpretations presented should be considered within the context of this particular application. While the results shed light on the factors and processes that may influence friendship networks among students, they may not be generalizable to other populations or settings. Therefore, caution should be exercised when extrapolating these findings to broader contexts.

References

Aldous, D. J. (1981). Representations for partially exchangeable arrays of random variables. *Journal of Multivariate Analysis, 11*(4), 581–598.

Amati, V., Schönenberger, F., & Snijders, T. A. (2015). Estimation of Stochastic actor-oriented models for the evolution of networks by generalized method of moments. *Journal de la Société Française de Statistique, 156*(3), 140–165.

Amati, V., Schönenberger, F., & Snijders, T. A. (2019). Contemporaneous statistics for estimation in stochastic actor-oriented co-evolution models. *Psychometrika, 84*(4), 1068–1096.

Amoroso, S., Coad, A., & Grassano, N. (2018). European R&D networks: A snapshot from the 7th EU Framework Programme. *Economics of Innovation and New Technology, 27*, 404–419. https://doi.org/10.1080/10438599.2017.1374037

Anderson, C. J., Wasserman, S., & Crouch, B. (1999). A p* primer: Logit models for social networks. *Social Networks, 21*(1), 37–66.

Barabási, A. L., & Albert, R. (1999). Emergence of scaling in random networks. *Science, 286*(5439), 509–512.

Barrios, C., Flores, E., & Martínez, M. Á. (2019). Club convergence in innovation activity across European regions. *Papers in Regional Science, 98*, 1545–1565. https://doi.org/10.1111/pirs.12429

Bauer, V., Harhoff, D., & Kauermann, G. (2022). A smooth dynamic network model for patent collaboration data. *AStA Advances in Statistical Analysis, 106*, 97–116. https://doi.org/10.1007/s10182-021-00393-w

Besag, J. E. (1972). Nearest-neighbour systems and the auto-logistic model for binary data. *Journal of the Royal Statistical Society: Series B (Methodological), 34*(1), 75–83.

Ceoldo, G., Snijders, T. A., & Wit, E. C. (2023). Stochastic actor oriented model with random effects. *arXiv preprint arXiv:2304.07312*.

Coleman, J. S. (1988). Social capital in the creation of human capital. *The American Journal of Sociology, 94*, S95–S120.

Daudin, J. J., Picard, F., & Robin, S. (2008). A mixture model for random graphs. *Statistics and Computing, 18*(2), 173–183.

Dekker, D., Krackhardt, D., & Snijders, T. A. (2007). Sensitivity of MRQAP tests to collinearity and autocorrelation conditions. *Psychometrika, 72*, 563–581.

Denny, M. J., et al. (2017). Gergm: Estimation and fit diagnostics for generalized exponential random graph models. *R Package Version, 0*(11), 2.
Desmarais, B. A., & Cranmer, S. J. (2012). Statistical inference for valued-edge networks: The generalized exponential random graph model. *PLoS One, 7*(1), e30136.
Erdős, P., & Rényi, A. (1963). Asymmetric graphs. *Acta Mathematica Academiae Scientiarum Hungarica, 14*(295–315), 3.
Filippetti, A., & Zinilli, A. (2023). The innovation networks of city-regions in Europe: Exclusive clubs or inclusive hubs? *Papers in Regional Science, 102*(6), 1169–1193.
Frank, O., & Strauss, D. (1986). Markov graphs. *Journal of the American Statistical Association, 81*, 832–842.
Geyer, C. J., & Thompson, E. A. (1992). Constrained Monte Carlo maximum likelihood calculations. *Journal of the Royal Statistical Society B, 54*, 657–699.
Gilbert, E. N. (1959). Random graphs. *The Annals of Mathematical Statistics, 30*(4), 1141–1144.
Glückler, J. (2007). Economic geography and the evolution of networks. *Journal of Economic Geography, 7*, 619–634.
Handcock, M. S., Robins, G., Snijders, T., Moody, J., & Besag, J. (2003). *Assessing degeneracy in statistical models of social networks (Vol. 39)*. Working paper.
Handcock, M. S., Hunter, D. R., Butts, C. T., Goodreau, S. M., & Morris, M. (2008). ergm: A package to fit, simulate and diagnose exponential-family models for networks. *Journal of Statistical Software, 24*(3) URL http://www.jstatsoft.org/v24/i03/
Handcock, M. S., Hunter, D. R., Butts, C. T., Goodreau, S. M., Krivitsky, P. N., & Morris, M. (2015). ergm: Fit, simulate and diagnose exponential-family models for networks. *The Statnet Project*, 3.5.1 edition.
Hanneke, S., Wenjie, F., & Xing, E. P. (2010). Discrete temporal models of social networks. *Electronic Journal of Statistics, 4*, 585–605. https://doi.org/10.1214/09-EJS548
Hawe, P., & Ghali, L. (2008). Use of social network analysis to map the social relationships of staff and teachers at school. *Health Education Research, 23*(1), 62–69.
Holland, P. W., & Leinhardt, S. (1981). An exponential family of probability distributions for directed graphs. *Journal of the American Statistical Association, 76*(373), 33–50.
Huisman, M., & Snijders, T. A. (2003). Statistical analysis of longitudinal network data with changing composition. *Sociological Methods & Research, 32*(2), 253–287.
Hunter, D. R. (2007). Curved exponential family models for social networks. *Social Networks, 29*(2), 216–230.
Hunter, D. R., & Handcock, M. S. (2006). Inference in curved exponential family models for networks. *Journal of Computational and Graphical Statistics, 15*(3), 565–583.
Hunter, D. R., Handcock, M. S., Butts, C. T., Goodreau, S. M., & Morris, M. (2008a). ergm: A package to fit, simulate and diagnose exponential-family models for networks. *Journal of Statistical Software, 24*(3), nihpa54860.
Hunter, D. R., Goodreau, S. M., & Handcock, M. S. (2008b). Goodness of fit of social network models. *Journal of the American Statistical Association, 103*(481), 248–258. https://doi.org/10.1198/016214507000000446
Krackardt, D. (1987). QAP partialling as a test of spuriousness. *Social Networks, 9*(2), 171–186.
Krivitsky, P. N. (2012). Exponential-family random graph models for valued networks. *Electronic Journal of Statistics, 6*, 1100.
Krivitsky, P. N., & Handcock, M. S. (2008). Fitting position latent cluster models for social networks with latentnet. *Journal of Statistical Software, 24*(5).
Krivitsky, P. N., & Handcock, M. S. (2014). A separable model for dynamic networks. *Journal of the Royal Statistical Society. Series B, Statistical Methodology, 76*(1), 29.
Krivitsky, P. N., Handcock, M. S., Hunter, D. R., & Krivitsky, M. P. N. (2012). Package 'ergm. count'. *Journal of Statistics, 6*, 1100–1128.
Krivitsky, P. N., Hunter, D. R., Morris, M., & Klumb, C. (2023). ergm 4: New features for analyzing exponential-family random graph models. *Journal of Statistical Software, 105*, 1–44.

Lazega, E., & Snijders, T. A. (Eds.). (2015). *Multilevel network analysis for the social sciences: Theory, methods and applications* (Vol. 12). Springer.

Liu, X., Derudder, B., & Liu, Y. (2015). Regional geographies of intercity corporate networks: The use of exponential random graph models to assess regional network-formation. *Papers in Regional Science, 94*, 109–126.

Long, E., Barrett, T., & Lockhart, G. (2019). Chronic health conditions and adolescent friendship: Perspectives from social network analysis. *International Journal of Adolescent Medicine and Health, 33*(5), 20180293.

Long, E., Gardani, M., McCann, M., Sweeting, H., Tranmer, M., & Moore, L. (2020). Mental health disorders and adolescent peer relationships. *Social Science & Medicine, 253*, 112973.

Lusher, D., Koskinen, J., & Robins, G. (Eds.). (2013). *Exponential random graph models for social networks: Theory, methods, and applications*. Cambridge University Press.

Metropolis, N., Rosenbluth, A. W., Rosenbluth, M. N., Teller, A. H., & Teller, E. (1953). Equation of state calculations by fast computing machines. *The Journal of Chemical Physics, 21*(6), 1087–1092.

Nowicki, K., & Snijders, T. A. B. (2001). Estimation and prediction for stochastic blockstructures. *Journal of the American Statistical Association, 96*(455), 1077–1087.

Pattison, P., & Robins, G. (2002). Neighborhood-based models for social networks. *Sociological Methodology, 32*(1), 301–337.

Pattison, P. E., & Wasserman, S. (1999). Logit models and logistic regressions for social networks, II. Multivariate relations. *British Journal of Mathematical and Statistical Psychology, 52*, 169–194.

Rinaldo, A., Fienberg, S. E., & Zhou, Y. (2009). *On the geometry of discrete exponential families with application to exponential random graph models*.

Ripley, R. M., Snijders, T. A., Boda, Z., Vörös, A., & Preciado, P. (2011). *Manual for RSIENA*. University of Oxford, Department of Statistics, Nuffield College, 1, 2011.

Robbins, H., & Monro, S. (1951). A stochastic approximation method. *The Annals of Mathematical Statistics, 22*, 400–407.

Robins, G. L., Pattison, P. E., & Wasserman, S. (1999). Logit models and logistic regressions for social networks. III. Valued relations. *Psychometrika, 64*, 371–394.

Robins, G., Pattison, P., Kalish, Y., & Lusher, D. (2007). An introduction to exponential random graph (p*) models for social networks. *Social Networks, 29*, 173–191.

Schweinberger, M. (2011). Instability, sensitivity, and degeneracy of discrete exponential families. *Journal of the American Statistical Association, 106*(496), 1361–1370.

Snijders, T. A. (1996). Stochastic actor-oriented models for network change. Journal of Mathematical Sociology, 21(1–2), 149–172.

Snijders, T. A. B. (2001). The statistical evaluation of social network dynamics. *Sociological Methodology, 31*(1), 361–395.

Snijders, T. A. (2002). Markov chain Monte Carlo estimation of exponential random graph models. *Journal of Social Structure, 3*(2), 1–40.

Snijders, T. A. (2017). Stochastic actor-oriented models for network dynamics. *Annual Review of Statistics and Its Application, 4*, 343–363.

Snijders, T. A., & Borgatti, S. P. (1999). Non-parametric standard errors and tests for network statistics. *Connect, 22*(2), 161–170.

Snijders, T. A., Pattison, P. E., Robins, G. L., & Handcock, M. S. (2006). New specifications for exponential random graph models. *Sociological Methodology, 36*(1), 99–153.

Snijders, T. A., Koskinen, J., & Schweinberger, M. (2010). Maximum likelihood estimation for social network dynamics. *The Annals of Applied Statistics, 4*(2), 567.

Steglich, C., Snijders, T. A., & West, P. (2006). Applying siena. *Methodology, 2*(1), 48–56.

Strauss, D., & Ikeda, M. (1990). Pseudolikelihood estimation for social networks. *Journal of the American Statistical Association, 85*(409), 204–212.

Watts, D. J. (2004). The 'new' science of networks. *Annual Review of Sociology, 30*, 243–270.

Watts, D. J., & Strogatz, S. H. (1998). Collective dynamics of 'small-world' networks. *Nature, 393*(6684), 440–442.
Wilson, J. D., Denny, M. J., Bhamidi, S., Cranmer, S. J., & Desmarais, B. A. (2017). Stochastic weighted graphs: Flexible model specification and simulation. *Social Networks, 49*, 37–47.
Yon, G. G. V., Slaughter, A., & de la Haye, K. (2021). Exponential random graph models for little networks. *Social Networks, 64*, 225–238.
Zinilli, A. (2016). Competitive project funding and dynamic complex networks: *Evidence from Projects of National Interest (PRIN) Scientometrics, 108*(2), 633–652. https://doi.org/10.1007/s11192-016-1976-4.
Zinilli, A., & De Marchi, M. (2020). Value-added in high technology and industrial basic research: A weighted network observing the trade of high-tech goods. *International Journal of Computational Economics and Econometrics, 10*(4), 398–418.
Zinilli, A., Yujie, G., & Scherngell, T. (2024a). Structural Dynamics of Inter-city Innovation Networks in China: A Perspective From TERGM. *Networks and Spatial Economics, 24*(3), 707–741. https://doi.org/10.1007/s11067-024-09634-2.
Zinilli, A., Pierucci, E., & Reale, E. (2024b). Organizational factors affecting higher education collaboration networks: Evidence from Europe. *Higher Education, 88*(1), 119–160. https://doi.org/10.1007/s10734-023-01109-6.

MIX
Papier aus verantwortungsvollen Quellen
Paper from responsible sources
FSC® C105338
FSC
www.fsc.org

If you have any concerns about our products,
you can contact us on
ProductSafety@springernature.com

In case Publisher is established outside the EU,
the EU authorized representative is:
Springer Nature Customer Service Center GmbH
Europaplatz 3, 69115 Heidelberg, Germany

Printed by Libri Plureos GmbH
in Hamburg, Germany